José Luis Rios Flores
Luis Gererdo Yáñez Chávez
Manuel de Jesus A. Ruiz-Esparza

Productivity of water, capital and labor in apple cultivation

José Luis Rios Flores
Luis Gererdo Yáñez Chávez
Manuel de Jesus A. Ruiz-Esparza

Productivity of water, capital and labor in apple cultivation

From Cuauhtémoc, Chihuahua, Mexico in producers with low, medium and high use of technology

ScienciaScripts

Imprint

Cover image: www.ingimage.com

This book is a translation from the original published under ISBN 978-620-2-15082-8.

Publisher:
Sciencia Scripts
is a trademark of
Dodo Books Indian Ocean Ltd. and OmniScriptum S.R.L publishing group

120 High Road, East Finchley, London, N2 9ED, United Kingdom
Str. Armeneasca 28/1, office 1, Chisinau MD-2012, Republic of Moldova, Europe
Printed at: see last page
ISBN: 978-620-7-67241-7

Contents

SUMMARY 2
CHAPTER I 3
CHAPTER II 5
CHAPTER III 7
CHAPTER IV 16
CHAPTER V 20
CHAPTER VI 45
LITERATURE CITED 48

SUMMARY

The objective was to determine the productivity of water, labour and capital in the cultivation of red apple (*Malus domestica*) in three different degrees of technification in Cuauhtemoc, Chihuahua, Mexico. Indicators that evaluated the physical, economic and social productivity of water (water footprint), labour and capital were determined by applying mathematical models of our own creation. The results indicate that Cuauhtemoc is the main producer of red apple in Mexico, contributing 25.95% of the US$4,265.3 million generated by the crop at the national level. Water productivity, in physical, economic and social terms, was as follows: physical, 0.91, 2.02 and 3.18 kg m^{-3} (the world average is 70 litres/100 gr of apple, i.e. 1.428 kg m^{-3}), economic, US$84,341, US$284,726 and US$614,244 of gross profit per hm^3 , and social, 28.1, 19.6 and 22.7 jobs generated per hm^3 of irrigated water for producers with low (BT), medium (MT) and high (AT) use of technology respectively, while for capital productivity had indicators of 49.4, 58.4 and 86.1% profit rate respectively for BT, MT and AT, finding that profitability was decreasing between 2002 and 2013 for AT (decreased its Benefit-Cost Ratio "RB/C" from 2.38 to 1.86) and MT (decreased its RB/C from 1.86 to 1.58) and increasing in BT (decreased its RB/C from 1.86 to 1.58).

(increasing its RB/C from 1.37 to 1.49), and the social productivity of capital was 164.5, 40.1 and 31.8 jobs per million US$ invested for LV, MV and HV respectively. Each hour of work produced in LV, MV and HV, in physical terms, 14.0, 44.8 and 60.8 kg, and US$1.30, US$6.31 and US$11.73 profit. The cost per kg was US$0.19 (BT), US$0.24 (MT) and US$0.22 (AT).

Keywords: Productivity, Imdrica footprint, social efficiency.

CHAPTER I

INTRODUCTION

Of all sectors of the economy, agriculture is the most sensitive to water scarcity, employing approximately 70% of global freshwater withdrawals, making it the sector with the most possibilities or options for adjustment (UN-Water, 2012)1. The Mdric Footprint assessment contributes with a new perspective, in which total water needs are quantified and geographically localised (Aldaya *et al.,* 2011)[1 2] . The concept of a metric footprint was introduced as an analogy to the ecological footprint, indicating water use rather than land use. The metric footprint is an indicator of water use that shows both direct and indirect use of the water resource. It is defined as the total volume of freshwater used to produce a given good or product. The metric footprint is a geographically explicit indicator that not only shows the volumes of water use and pollution, but also the locations (Hoekstra *et al.,* 2011)[3] .

Three colours are distinguished in the metric footprint: blue, green and grey. The blue metric footprint refers to the consumption of water resources from surface water and groundwater. Consumption" is understood as the loss of water from a body of surface water or groundwater in a river basin as the water evaporates, returns to another basin, to the sea or is incorporated into a product. The green water footprint refers to the consumption of rainwater stored in the soil as moisture (Falkenmark and Rockstrom, 20044; 20065). The grey water footprint is a concept still under debate and refers to pollution. It is defined as the volume of freshwater needed to assimilate a load of pollutants based on current environmental water quality standards (Samoral *et al.,* 2012)[4 5 6] .

---

[1] UN-Water, (2012). The United Nations World Water Development: Managing Water under uncertainty and risk. World Water Assessment Programmed (WWAP). Report 4. Unesco, Paris, France.861p.

[2] Aldaya M. M., Niemeyer I, and Zarate E. (2011). Water and Globalisation: Challenges and opportunities for a better management of water resources. Revista Espanola de Estudios Agrosociales y Pesqueros, 2011; 230: 61-83.

[3] Hoekstra Arjen Y., Ashok K. Chapagain, Maite M. Aldaya and Mesfin M. Mekonnen. Chapagain, Maite M. Aldaya and Mesfin M. Mekonnen (2011). The Water Footprint Assessment Manual Setting the Global Standard. Earthscan Ltd, Dunstan House, 14a St Cross Street, London. ISBN: 978-1-84971-279-8.

[4] Falkenmark, M., & Rockstrom, J. (2004). *Balancing water for humans and nature: the new approach in ecohydrology*. Earthscan.

[5] Falkenmark, M. & Rockstrom, J. (2006). The New Blue and Green Water Paradigm: Breaking New Ground for Water Resources Planning and Management. *J. Water Resour. Plann. Manage*. 132(3), 129-132.

[6] Salmoral, G., Dumont, A., Aldaya, M. M., Rodriguez-Casado, R., Garrido, A., & Llamas, M. R. (2012). *Analysis of the extended Ivdric footprint of the Guadalquivir river basin*. Marcelino Botin Foundation.

In crop production, the objective of promoting efficient water use is to produce higher economic yields with less water when water is a limiting factor, as in arid zones (Boutraa, 2010)[7] . Therefore, it is urgent to develop appropriate strategies for saving and conserving groundwater (blue gold) to promote sustainable agricultural and human systems, which can be achieved through irrigation systems appropriate to the current situation (Postel, 2000)[8] . Apple trees require irrigation to achieve high yields and fruit quality, but as water supply has become a limiting factor in the main production areas, the exploitation of this crop must be oriented in terms of water productivity (Passioura, 2006)[9] . Therefore, the objective of this work was to determine the water footprint of apple tree production in the municipality of Canatlan, Durango, of low and high technology producers.

[7] Boutraa, T. (2010). Improvement of water use efficiency in irrigated agriculture: a review. *Journal of Agronomy*, *9*(1), 1-8.
[8] Postel, S. L. (2000). Entering an era of water scarcity: the challenges ahead. *Ecological applications*, *10*(4), 941-948.
[9] Passioura, J. (2006). Increasing crop productivity when water is scarce-from breeding to field management. *Agricultural water management*, *80*(1), 176-196.

CHAPTER II

II. PROBLEM TO BE ADDRESSED, OBJECTIVE AND HYPOTHESIS

2.1 Problem to be addressed and objective

The amount of fresh water on the planet does not exceed 3% of the total amount of water on the planet, which in absolute terms amounts to 35 million km^3 . Moreover, a million years ago, it was estimated that the earth had just under one million inhabitants, but today there are already 7.2 billion human beings on the planet.[3] 2 billion human beings on the planet, which means that, whereas a million years ago, there was no pressure on freshwater resources, as each human corresponds to 35 km of freshwater (UN, 2014)[10] . Currently, each human being only accounts for 0.00486 km^3 , while one m^3 , of water generates $512 pesos of profit when used in industry, in agriculture on average only $2 is generated, in addition to the high risk associated with agricultural production, on the other hand, agriculture is the macro user that consumes the most fresh water: 82% on average worldwide, although in some countries, such as the USA, agriculture accounts for only 40% of total freshwater consumption, while in backward countries agriculture consumes up to 95% of freshwater, the difference being used by industry and the domestic sector, as well as energy generation. Water is an extremely important resource for mankind, without it, life would simply not be possible for the human species. However, it seems that humanity has not become clearly aware of its importance, and acts as if water were an infinite resource, wasting it, polluting it and making inefficient use of it (Gamez, 2014)11.[11]

It is in this sense that the present work is framed in the *problem* of the efficient and productive use of the very *scarce water* resource by the agricultural sector, having as general *objective* the generation of numerical indicators that reflect the amount of irrigation water used in the production of a kilogram of physical product, as well as, of the amount of monetary value that is

[10] UN, 2014. United Nations. Department of Economic and Social Affairs, http://www.un.org/es/html

[1111] Gamez, A.B. 2014. Strawberry *(Fragaria vesca)* production of DR-14 Rio Colorado, BC and DR-085 Celaya, Guanajuato and its water footprint. Bachelor's thesis. Universidad Autonoma Chapingo. Unidad Regional Universitaria de Zonas Aridas, Bermejillo, Durango.

generated when irrigating a m^3 of water. While labour and capital factors are finite, they are not as scarce as water, but this does not preclude the need for indicators of the most productive use of these factors. On the other hand, there is little or no information on the productivity of water, labour and capital in the production of red apple (*Malus domestica*) in Mexico, specifically in the municipality of Cuauhtemoc, Chihuahua, Mexico, so the *specific objective* was to determine numerous indicators that measure the productivity of water, labour and capital in the cultivation of apple trees, for three different levels of production technology.

2.2 Hypotheses

First hypothesis: ***Irrigation water*** productivity in red apple (*Malus domestica)* cultivation in Cuauhtemoc, Chihuahua (measured as water footprint as **L $_{kg-1}$, and kg $_{m-3}$ and profit and employment generated per hm^3**) is ***higher*** in highly technified "AT" producers than in producers with medium "MT" and low "BT" use of technology.

Second hypothesis: The productivity of ***capital*** in the cultivation of red apple (*Malus domestica)* in Cuauhtemoc, Chihuahua (measured as productivity in the form of **Benefit Cost Ratio and employment generated per million dollars invested in production**) is ***higher*** in highly technified producers "AT" in relation to producers of medium (MT) and low (BT) use of technology.

Third hypothesis: The productivity of ***labour*** in the cultivation of red apple (*Malus domestica)* in Cuauhtemoc, Chihuahua (measured in the form of **kg of apple and**

US$ of profit produced per hour of work) is ***higher*** in highly "A" technified producers in relation to medium (MT) and low (BT) technified producers.

CHAPTER III

III. LITERATURE REVIEW

3.1 The water-food connection

Water has a great diversity of roles on Planet Earth, being required in all terrestrial ecosystems from which we derive our food. Water is directly related to man's food needs as an essential food, as they cannot survive without drinking water for more than a few days. The irony is that while drinking water needs are a few litres a day, the water incorporated into the human daily diet can be a thousand times more. The lack of public awareness of this fact is just one example of the limited understanding and visibility that most water issues have in modern societies.

Population growth and economic development in recent decades have driven the demand for food and thus the demand for water. At the same time as the demand for water for agriculture increased, other sectors of society also increased their demands, in particular the environment. Given the economic and environmental constraints to increasing water supply to meet growing demand, the prospects for water scarcity are increasing in many regions of the world (Molden, 2007)[13] . Among the many dimensions of the food security issue, the availability of water for food production has become the most critical. For example, there is a perception among water experts that increasing demands from other sectors of society could, in the future, limit water allocation for irrigated agriculture (Jurado and Vaux, 2005)[14] .

The water required to meet food needs is substantial and is driven inexorably upwards by population growth and also by economic development. Estimates from Harrison *et al.,* (2002)[15] indicate that the water footprint has increased on average 20%-30% over the time in which modern economic development took place, derived from increased consumption of animal proteins and fats. Therefore, even if population growth is checked, the per capita increase in water demand for food due to economic development is also

[13] Molden D. 2007. *Water for food, water for life: a comprehensive assessment of water management in agriculture. London: Earthscan; Colombo: International Water Management Institute; 2007.*

[14] Jury, W. A., & Vaux, H. (2005). The role of science in solving the world's emerging water problems. *Proceedings of the national academy of sciences of the united states of america, 102*(44), 15715-15720.

[15] Harrison, P., Bruinsma, J., de Haen, H., Alexandratos, N., Schmidhuber, J., Bodeker, G., & Ottaviani, M. G. (2002). World agriculture: towards 2015/2030. *Online, http://www. fao. org/documents.*

inevitable, unless drastic changes in diet occur in the future.

Following the rapid increase in crop productivity worldwide between 1960 and 1990, policy makers in science became convinced that the Earth's capacity to produce food was large enough to meet any future demand. This led to a decline in agricultural research efforts, with the exception of investments in major research on plant molecular biology and climate change. The few early warnings regarding the uncertainties faced in meeting the challenge of food security (Penning de Vries, 2001[16] , on land degradation) have been followed more recently by a renewed interest in this topic, especially as a consequence of the sharp increase in food prices in 2008. Subsequent fluctuations in world prices show a steady upward trend that undoubtedly has many drivers (misdirected policies, speculation, competition from biofuel production, etc.), but is also affected by a fundamental imbalance between supply and demand. It is therefore not surprising that food security has not only moved to the forefront of agricultural research, but is now perceived as an important issue for more fundamental research (Nature, 2010)[17] .

3.1.1 Water Scarcity and Food Production

According to Fereres *et al.,* (2011)[18] food security has many dimensions, but, in terms of water availability, the main question is ^Will there be enough food in the foreseeable future? and ^Will there be enough water to produce enough food? There are basically three strategies to cope with water scarcity in food production.

- Increase water supply and land above current levels.
- Increase water productivity, either by improving yield or by improving water use efficiency, or both.
- Importing water in the form of food (virtual water), an option to address regional water scarcity, which depends mainly on the global economy and trade.

The development of new water sources and/or the reallocation of water from other sectors to increase food production has limited potential in some areas and is no longer possible in other regions of

---

16 De Vries, F. P. (2001). Food Security? We Are Losing 1 Ground Fast! *Crop science: Progress and prospects*, 1.

17 Nature. Can science feed the world? 2010. http://www.nature.com/news/specials/food/index.html. Access July 30 from 2015.

18 Fereres, E., Orgaz, F., & Gonzalez-Dugo, V. (2011). Reflections on food security under water scarcity. *Journal of Experimental Botany*, *62*(12), 4079-4086.

the world. Furthermore, if reallocation is done at the expense of the environment, this could lead to the degradation of environmental services and cause irreversible negative impacts on ecosystems. Agricultural land expansion is another option that has been responsible for 20% of the increase in food production over the last decades, but has limited potential due to urbanisation and soil degradation processes (De Vries, 2001)[19] .

The strategy of increasing productivity is primarily responsible for the 80% increase in production, but this increase in production has not been achieved without negative impacts on the environment in ways that threaten the sustainability of farming systems. There is no doubt that increasing productivity is the main strategy for the future, but at the same time as water resources are limited, farming systems will have to become more sustainable. Productivity increases have arisen mainly as a consequence of yield increases. (Fischer *et al.,* 2009)19 and much less due to the reduction of water use.

3.2 Rainfed agriculture and food production

It is important to emphasise that the boundaries between rainfed and irrigated agriculture are becoming blurred. On the one hand, uncertainty in water supply to many irrigated areas generates variations in irrigation versus rainfed areas each season. On the other hand, rainfed farmers in many areas of the world are willing to develop storage facilities to provide supplemental irrigation to their crops when possible. On the other hand, in most areas, rainfall is an important part of the seasonal consumptive use of irrigated crops, so the fashionable trend of separating "blue" from "green" water (Falkenmark and Rockstrom, 2006) is questionable.

Clearly, only a small proportion of the world's cultivated area is equipped for irrigation (between 15-20%), and important economic, environmental and institutional constraints restrict the expansion of irrigation. In fact, irrigated agriculture at the regional scale will be very dynamic in the future, expanding in some areas and declining in others, constrained by water availability, salinity, and markets. Increasing the productivity of rainfed systems will be crucial, as they

[19] De Vries, F. P. (2001). Food Security? We Are Losing 1 Ground Fast! *Crop science: Progress and prospects*, 1.
[19] Fischer, R. A., Byerlee, D., & Edmeades, G. O. Can technology deliver on the yield challenge to 2050? 46p

cover the majority of agricultural land. Rainfall variability is the main source of uncertainty and the main cause of large fluctuations in farmers' incomes. The threat that low-income years pose to the sustainability of agriculture explains the conservative nature of most farm management practices adopted by rainfed farmers. However, practices aimed at avoiding extreme drought years tend to leave resources unused in good years because they do not take full advantage of periodically favourable conditions. In many rainfed systems, the balance between productivity and sustainability must be tilted more towards productivity in the future, either by accepting more risk or decreasing risk. To decrease risk, attention should focus on increasing crop water supply, either by using conservation tillage where appropriate, or by developing supplementary irrigation where feasible. The best option to reduce risk is to manage prudently, based on a number of tools and approaches being developed to provide pre-season rainfall predictions and expected outcomes of management decisions (Cooper *et al.,* 2008)20. [20]This is an area of research that deserves much more attention, as it would lead to rainfed agriculture becoming more sustainable.

3.3 Irrigated agriculture and water productivity

There is no doubt that irrigated agriculture is essential to meet future food demand, but it also faces numerous problems that threaten its sustainability. The problem of salinity, drainage is inherent to irrigated agriculture and cannot be solved permanently, but long-term research is needed to optimise salinity management and to minimise the environmental impact of irrigation. Future water shortages will probably lead to a widespread use of deficit irrigation, a practice that increases the efficiency of irrigation water use but requires precise control of both water stress levels and salinity build-up. Another major threat to irrigation sustainability is the overexploitation of groundwater, a growing problem in many areas of the world. For example, in the North China Plain, the consumptive use of the current wheat-maize rotation far exceeds the sustainable recharge of the area's aquifers, causing the

[2020] Cooper, P. J. M., Dimes, J., Rao, K. P. C., Shapiro, B., Shiferaw, B., & Twomlow, S. (2008). Coping better with current climatic variability in the rain-fed farming systems of sub-Saharan Africa: An essential first step in adapting to future climate change? *Agriculture, Ecosystems & Environment*, *126*(1), 24-35.

groundwater table to decline steadily over the last decade.

3.4Irrigated agriculture linked to overexploitation of aquifers in Mexico

In Mexico, in 2010, about 37% of the total concessioned volume for consumptive uses, i.e. uses that consume water in the activity itself, came from groundwater (CONAGUA, 2011a)[21] . The main grouped water use in Mexico, agriculture -which includes agricultural, aquaculture, livestock, multiple and other uses- has a concessioned volume of around 61.8 billion m^3 $year^{-1}$, of which 33.8% is extracted from groundwater (CONAGUA, 2011b)[22] . On the other hand, at the national level, of the 11.4 billion m3/year concessioned for urban and domestic public use, 62.2% comes from groundwater.

According to CONAGUA, (2011b) of the 653 aquifers in Mexico, 32 were overexploited in 1975, i.e. in 32 aquifers more water was being extracted than was being recharged. In 1985 there were already 80 aquifers and by 31 December 2009 there were 100 aquifers, from which 53.6% of the national groundwater was extracted. It is a worrying fact that more than half of the groundwater comes from overexploited aquifers, but it is even more alarming that according to CONAGUA (2011a), the pooled agricultural use of groundwater has increased to 23.2% from 2001 to 2009 and, furthermore, that urban and domestic public uses of groundwater have grown by 30.3% in the same period. Of the one hundred aquifers already overexploited at the national level in 2009, 72 are located in the states of Sonora, Chihuahua, Baja California, Baja California Sur, Coahuila, Durango, Nuevo Leon, Guanajuato, Puebla, San Luis Potosi, Zacatecas, Estado de Mexico and Queretaro (CONAGUA, 2011b).

The problem of overexploitation of water is widespread, in Mexico Caravanes (2013)[23] , reports that of the 314 292 water rights, 76% of the concessioned volume is extracted for agricultural purposes, 16% for urban public use and 5.2% for industrial use. Most of the wells are divided between agricultural and urban public use. The

---

[21] CONAGUA (2011a), Atlas del Agua en Mexico, SEMARNAT-Gobierno Federal, Mexico.

[22] CONAGUA (2011b), Estadisticas del Agua en Mexico, National Water Commission. Available at: www.conagua.gob.mx

[23] Caravantes, R. E. D., Pena, L. C. B., Cejudo, L. C. A., & Flores, E. S. (2013). Anthropogenic pressure on groundwater in Mexico: a geographic approach. *Investigaciones Geograficas, Boletm del Instituto de Geografia*, *2013*(82), 93-103.

same study shows that the areas with the highest pressure are distributed throughout the Mexican territory. However, all the figures indicate a

The total pressure is very strong in the Bajfo region, the Comarca Lagunera and central-western Chihuahua. These findings are discussed from a geographical perspective, demonstrating the usefulness of incorporating environmental indicators to assess the pressure exerted on the country's water resources (Fig. 1).

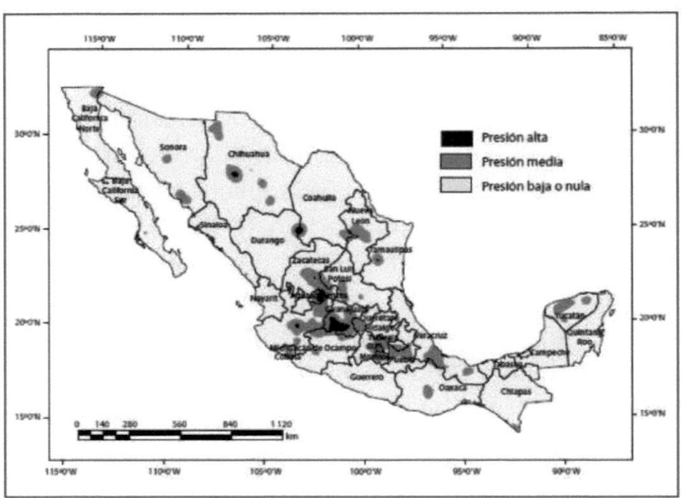

Fig. 1. Areas of anthropogenic pressure on groundwater in Mexico. Source Caravanes *et al.,* (2013).

3.4.1 Cuauhtemoc case, Chihuahua

Chihuahua is the largest state in the country, representing one eighth of the total surface of the national territory (Plan Estatal de Desarrollo 2004- 2011, 2004)[24] . A large part of this territory is made up of agricultural areas, which are the main economic activity of the state. The region of Cuauhtemoc is one of the most important in the state, as it produces a variety of crops such as corn, beans and oats, mainly. Likewise, the state of Chihuahua is the main producer of apples in the country, with a contribution of more than 58% of the national production, in addition, more than 75% of the national

[24] State Development Plan 2004-2010. Chihuahua State Government. pp. 99-101.

product is selected and packed in the state (UACH, 2007)[25] .
According to the State Development Plan of the Chihuahua State Government (2006)[26] , agriculture is supplied with more than 85% of the water used in the state, through more than 13500 wells registered in the state, with which approximately 338 thousand hectares are irrigated. Of these wells, 3650 are located in the Cuauhtemoc region, framed by the Bustillos Lagoon Basin. But due to excess extraction and low efficiency in water conduction, as well as climatic factors, the vast majority of the state's aquifers are going through a stage of overexploitation, as is the case of the Cuauhtemoc Aquifer.

A fundamental characteristic of the area is that the aquifer is recharged only by rainwater, with an average annual recharge of only 115 hm^3 (CONAGUA and COLPOS, 2007)27. If this is associated with the great drought that has been occurring in recent decades, it is possible to foresee the possible effects of overexploitation of the aquifer, especially if we consider that 569.4 hm^3 (CONAGUA, 2009)[27 28] are extracted annually, of which 92.7% are for agricultural use, 4% for the sustenance of the urban area and 3.3% for livestock activities.

3.5Water and apple tree cultivation in Cuauhtemoc, Chihuahua

The apple tree [Malus sylvestris (L.) Mill. var. Domestica (Borkh.) Mansf.], occupies one of the first places in the state of Chihuahua by the planted surface, production and quality of the fruit, being the Golden Delicious cultivar the most important (Guerrero-Prieto *et al.*, 2004)[29] . According to Pena and Alatorre, (2013)[30] , in the study area a total of 1467 wells were located within the cultivated agricultural category, and when summing up the volumes of registered concessioned volumes in the study area, the total number of wells in the cultivated agricultural category is

[25] UACH, (2007). Evaluacion de Alianza para el Campo de los Sistemas Producto Fruticolas en el Estado de Chihuahua. Autonomous University of Chihuahua, 26-37pp.

[26] State Development Plan 2010 -2016 (2010). Government of the State of Chihuahua, pp. 160-167.

[27] CONAGUA and COLPOS. (2007). Plan director: union de asociaciones de usuarios de aguas subterránraneas del acuifero de Cuauhtemoc, Chihuahua, S de RL de IP de CV, Comision Nacional del Agua.

[28] CONAGUA (2009). Actualizacion de la Disponibilidad Media Anual de Agua Subterranea Acuifero (0805) Cuauhtemoc, Estado De Chihuahua, Comision Nacional del Agua. Available at: http://www.conagua.gob.mx

[29] Guerrero-Prieto, V., Trevizo, G., Figueroa, C., Romo, A., Gardea, A., & Blanco, C. (2004). Identification of epiphytic yeasts obtained from apple [Malus sylvestris (L.) Mill. var. domestica (Borkh.) Mansf.] for postharvest biological control. *Revista Mexicana de Fitopatologia*, *22*(2), 223-230.

[30] Pena, A. G., & Alatorre, L. C. Evaluation of groundwater abstractions by indirect methods in the Cuauhtemoc region, Chihuahua, Mexico: applying remote sensing and GIS.

of the concessioned volumes registered in REPDA gives an extraction volume of 135.45 hm^3 (CONAGUA, 2010)[31] . This information shows that all the crops analysed (maize, oats, beans and apple trees) do not comply with this concessioned volume, which could be causing overexploitation of the aquifer. Based on Table 1, it can be said that all crops are exceeding the concessioned volume, especially corn. In the long term, this could cause the Cuauhtemoc Aquifer to present a serious decrease in its static level, causing not only problems for the continuation of agricultural activity, but also causing a shortage of water for the population of the region. In the Cuauhtemoc aquifer, the agricultural zone has an estimated extension of 54,568 hectares. Along this surface, the existing dynamics has caused that the exploitation of the water resources is considerably affected, more specifically, 569.4 hm^3 are extracted annually, distributed in 92.7% in agricultural use, 4% for urban public use, while 3.These actions have led to considerable overexploitation of the aquifer, especially if we take into account that the natural annual recharge of the aquifer is 115 hm^3 (CONAGUA, 2009)[32] .

Table. 1: Differences between concessioned and extracted volume in the Cuauhtemoc coal mine, Chihuahua 2011.			
Cultivation	**Vol. Concessioned (hm)^3**	**Vol. Required per agricultural cycle (hm)^3**	**Difference between Vol. required and concessioned.**
Oats	135.449	257.961	122.512
Ma^z	135.449	335.349	199.900
Bean	135.449	206.369	70.92
Apple	135.449	167.674	30.225
Figures calculated for the cultivated area of 51 for the 2011 agricultural cycle.			579.34 hectares

On the other hand, if we compare the results obtained with the studies carried out by the National Water Commission and the

[31] CONAGUA (2010). Registro Publico de Derechos de Agua, Comision Nacional del Agua. Retrieved October 15, 2010. Available at www.conagua.gob.mx

[32] CONAGUA (2009). Actualizacion de la Disponibilidad Media Anual de Agua Subterranea Acuifero (0805) Cuauhtemoc, Estado De Chihuahua, Comision Nacional del Agua. Available at: http://www.conagua.gob.mx

College of Postgraduates (CONAGUA and COLPOS, 2007)[33] , the situation is aggravating and confirms the predictions made for the Cuauhtemoc, Chih. aquifer, where the enormous recharge deficit leads us to estimate that in a period of no more than 12-15 years the aquifer will not be able to maintain this extraction (CONAGUA and COLPOS, 2007)[34] , with the consequent collapse of agricultural activities in the region, but on the other hand, the productive inertia, the enormous investment in agro-industrial infrastructure and the social benefits derived from it, form a very solid dam, form a very solid dike that prevents both the authorities of the three levels of government and the existing organisations themselves from starting to take the necessary corrective and preventive decisions, despite the fact that all the entities involved, public, private and civil associations, are fully aware that the future viability of the region is seriously threatened.

[33] CONAGUA and COLPOS. (2007). Plan director: union de asociaciones de usuarios de aguas subterránraneas del acuifero de Cuauhtemoc, Chihuahua, S de RL de IP de CV, Comision Nacional del Agua.
[34] CONAGUA and COLPOS. (2007). Plan director: union de asociaciones de usuarios de aguas subterránraneas del acuifero de Cuauhtemoc, Chihuahua, S de RL de IP de CV, Comision Nacional del Agua.

CHAPTER IV

IV. MATERIALS AND METHODS

4.1 Location of the study area

The municipality of Cuauhtemoc, Chihuahua is located in the central zone of the state at latitude 28° 25' 00", longitude 106° 52' 00" of the Grengwich Meridian, at an altitude of 2,010 metres above sea level. It is bordered to the north by the municipality of Namiquipa, to the east by Rivapalacio, to the south by Gran Morelos and Cusihuiriachi, and to the west by Bachiniva. Its climate can vary from semi-humid to temperate, with an annual average temperature of 14 °C. Its maximum temperature reaches 37.7° and its minimum is -14.6°C. It has an average annual rainfall of 66 days with a relative humidity of 65% and an annual rainfall of 439 mm and its dominant wind is from the southwest (Mata, 2004)[35] .

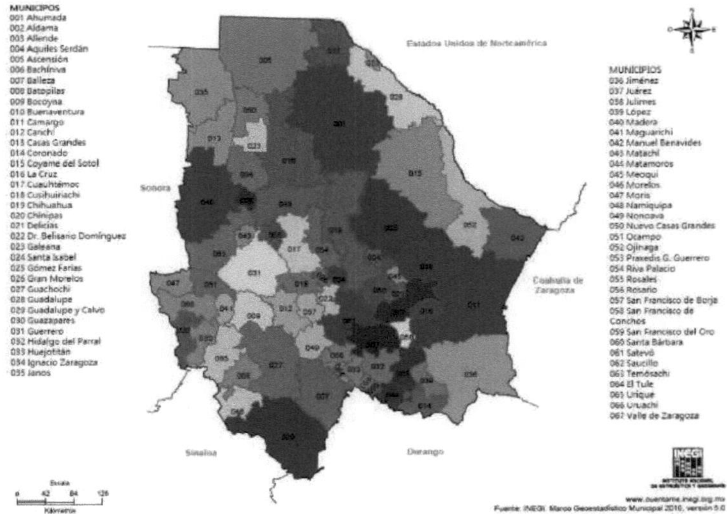

**Fig. 2. Geographical location of the municipality of Cuauhtemoc,
Chihuahua. Source INEGI (2010)[36]**

4.2 Sources of information

The information on production costs per hectare for low (BT),

[35] Mata, E., O. A. 2004. Population dynamics and biological management of nematodes associated with apple (Pyrus malus L.) in two localities in Cuauhtemoc, Chihuahua. Bachelor Thesis. Universidad Autonoma Agraria "Antonio Narro" Agronomy Division. Buenavista, Saltillo, Coahuila. Mexico. March 2004

medium (MT) and high (AT) technology use apple producers provided by UNIFRUT, based in Cuauhtemoc, Chihuahua, was used. The costs provided by UNIFRU also contain the average annual price structure (valued at market prices - producer prices - farm-gate prices) per tonne at which the[36] INEGI sold. 2010. Municipal Geostatistical Framework 2010, version 5.0.

BT, MT and AT producers, as well as the number of labour days invested per hectare in each cost component item, as well as the physical yield per hectare for AT and BT, only for MT producers were considered the physical yields and the average annual prices (valued at market prices - producer-to-farm prices) recorded by SIAP. Based on the Mexican peso-US dollar exchange rate issued by the Bank of Mexico on 14 April 2015 at 1256 hours of $15.365 pesos per dollar, all monetary variables were determined.

The monetary yield "RM" or revenue per hectare (from multiplying the physical yield per hectare by the price per tonne), in USD, is subtracted from the cost per hectare "C" and the profit per hectare is obtained. The profitability of the crop is then determined using the formula of the Benefit-Cost Ratio "BR/C", i.e.: BR/C = RM/C.

The volume "V" of water used in a hectare is the volume indicated by UNIFRUT in the calculation of the production costs per hectare, equal to 11,000 m^3 ha^{-1} , which is the amount of cubic metres that must be irrigated per hectare of red apple at commercial level.

A job was defined as the equivalent of the number of working days an average human being works in a year under average conditions. Thus, it was assumed that a person worked one day a day (a day being eight hours of work) for six days a week for 48 weeks a year, i.e. 6 days for 48 weeks = 288 days = 1 job.

4.3Variables evaluated and mathematical models used

1) Of land productivity, of general domain:

a) Physical yield "RF", measured in ton hectare-1, determined by the equation:

$$RF = \frac{\text{Quantity of physical product}}{\text{hectarea}}$$

b) Income per ha, also called monetary yield "RM", measured in US$, determined by the equation:

$$RM = \frac{RF * p}{una\ ha}$$

Where p = price ton^{-1} , in US$

c) Profit per hectare "g", measured in US$, determined by the equation:

$$g = \frac{RM - c}{una\ ha}$$

Where "c" is the cost per ha in US$.

2) Water productivity:

a) Kg m^{-3} . Determined by the equation

$$kg\,m^{-3} = \frac{RF}{V} = \frac{RF}{10{,}000LR} = 0.0001RF(LR)^{-1}$$

Where "V" is the volume of water used per hectare (in m^3), equivalent to the product of the irrigation rate "LR" per 10,000 m2.

b) m^3 kg^{-1} . Determined by the equation

$$m^3\,kg^{-1} = \frac{V}{RF} = \frac{10{,}000LR}{RF} = 10{,}000LR(RF)^{-1}$$

c) US$ gain hm^3 . Determined by the equation

$$\text{US\$ of profithm}^{-3} = 1{,}000{,}000\frac{g}{V}$$

d) Employment hm^3 . Determined by the equation

$$\text{Ernpleosge пeradospor hm}^3 = 1{,}000{,}000 * \left(\frac{J/288}{V}\right) = \frac{31250}{9}JV^{-1}$$

Where "J" is the number of working days per ha and 288 is the number of working days a worker has in a year at the rate of 6 working days per week for 48 weeks per year.

e) Water price per m^3 (in US$). Determined by the equation

$$\text{Water price per m}^3\ \text{Precio del agua por m}^3 = \frac{Cagua}{V}$$

Where "Cagua" is the cost of the item "Cost of water" within the Total cost per ha.

3) Capital productivity:

a) Benefit-Cost Ratio "RB/C". Determined by the Equation

$$RB/C = \frac{RM}{c}$$

b) Rate of gain "Tg". Determined by the equation

$$Tg = \frac{g}{c} * 100$$

c) social productivity of capital "E MUSD" (Jobs generated per million dollars invested in production). Determined by the equation

$$E\ MUSD = \frac{31250}{9}\left(\frac{J}{c}\right)$$

(b) Break-even point "EP". Determined by the equation

$$PE = \frac{c}{p}$$

4) Labour productivity:

b) ton hours^{-1} , determined by the equation

$$h\ ton^{-1} = \frac{J * 8}{RF}$$

c) kg h^{-1} , determined by the equation

$$kg\ h^{-1} = \frac{1000RF}{J * 8} = 125\frac{RF}{J}$$

d) US\$ of profit per working day, determined by the equation

$$US\$\ per\ day = \frac{g}{J}$$

e) US\$ profit per hour of work, determined by the equation

$$USD\ profit\ per\ hour = \frac{g}{J * 8}$$

CHAPTER V

V. RESULTS AND DISCUSSION

5.1 Fruit growing in the state of Chihuahua[39], its characteristics

The Union Agricola Regional de Fruticultores del Estado de Chihuahua (UNIFRUT), the source of the cost data for this study, as well as the number of day labourers per hectare, was founded in 1965 and is currently made up of 17 associations. UNIFRUT brings together fruit growers in Chihuahua, of which apple trees are the most important, although there are small, almost marginal areas of other fruit trees such as pistachio.

The 17 associations that make up UNIFRUT are the following:

1. Fruit Growers' Association of Alvaro Obregon
2. Fruit Growers Association of Bachiniva
3. Association of Fruit Growers of Basuchil
4. Association of Fruit Growers of Carchi
5. Association of Fruit Growers of Coyachi
6. Cuauhtemoc Fruit Growers' Associations
7. Cusihuiriachi Fruit Growers' Association
8. Association of Fruit Growers of Guerrero
9. Fruit Growers Association of Ignacio Zaragoza
10. Association of Fruit Growers of Las Cruces Namiquipa
11. Association of Fruit Growers of La junta
12. Association of Fruit Growers of Matachi
13. Association of Fruit Growers of Maguarichi
14. Association of Fruit Growers of Namiquipa
15. Association of Fruit Growers of Nvo. Casas Grandes
16. Temechi Fruit Growers' Association
17. Yepomera Fruit Growers' Association

The apple growing region is located in the northwest of the State of Chihuahua, it is the most important at national level; since the production of this region is 19.0 million boxes per year, equivalent to 78% of the national production. It is made up of around 2,500

[39] The information in this section was obtained through direct interview with UNIFRUT's credit management in Cuauhtemoc, Chihuahua.

producers, with an area of 30,000 hectares of apple trees in different stages of development.

That is to say, in the State of Chihuahua an average of 380,000 tons are produced per year, of which 70% is destined for fresh consumption and 30% for industry. The economic benefit of apple tree cultivation is of great importance, since 8 million daily workers are employed per year.

Currently, in the Regional Agricultural Union of Fruit Growers of the State of Chihuahua there are around 2,500 apple producers. Of these, 55% have a medium technological level. They are distinguished by the following characteristics:

- They do not have access to the latest technology and if they do have access, they do not know how to operate it in the most appropriate way to obtain the best benefits.
- Technical assistance is available, although in most cases it is not specialised.
- They have a technology package, but hardly ever apply it.
- They have micro-sprinkler irrigation systems and very few have more advanced irrigation systems, such as drip irrigation. They do not have re-pumping; therefore, irrigation intervals are very long and irrigation systems are operated empkically.
- Frost control methods are often deficient; therefore, damage can occur due to deficiencies in frost control.
- Therefore, they have medium yields ranging from 15.0 to a maximum of 25.0 tonnes per hectare.
- The quality of the apples is medium to low; therefore, they have difficulties in marketing the product.
- A very low percentage of the production is given added value (packaging).
- They are rarely creditworthy, as they do not have good credit histories and do not have sufficient collateral.
- The benefit-cost ratio for these producers is 1:2, i.e. for every peso invested, they get back two. The cost of production per hectare ranges between $66,000 pesos and yields per hectare range between 30.0 tons per hectare. On average, the production cost of one kilogram of apples is $2.20 pesos.

• To this type of producers, we can make them subject to Refaccionarios and Avfo credits, as long as they have technical support from an advisor.

31% are producing at a low technological level, distinguished by the following characteristics:

• They do not have technical assistance due to the lack of economic resources and the profitability of the production units.

• They are not creditworthy; therefore, it is almost impossible to have access to technologies for apple tree production.

• For the same reason, they do not have a Technology Package and even if in some cases they had one, it would be impossible to carry it out due to the lack of adequate tools.

• They do not have systems to prevent inclement weather, such as hail and frost. Therefore, yields are very low, ranging from 6.0 to a maximum of 15 tonnes per hectare.

• Productions tend to be of low quality; therefore, market prices for their products are often very low.

• The benefit-cost ratio for this type of producer is 1:0.50, i.e. for every peso invested, 50 cents are obtained. The cost of production per hectare ranges between $31,000 pesos and yields are between 11.0 tons per hectare. On average, the production cost per kilogram of apples is around $2.80 pesos.

• This type of producer can be brought to mid-tech with good planning, using financing as the main tool.

And only 14% of the producers are producing at a high technological level and are characterised by the following conditions:

They have state-of-the-art technology and know how to operate it to their advantage.

They have high-tech irrigation and re-pumping; therefore, irrigation intervals are not so long.

They have specialised technical assistance.

It has a technology package and usually carries it out.

Therefore, they have high yields of very high quality. The yield of this group of producers is around 60.0 tons per hectare or more. Their products are highly accepted in the markets; they usually give them added value (sorting, packaging, industrialisation). Its orchards have protection against bad weather, such as hail. It has good frost control, as it has the appropriate technology, such as: fans, heaters, central heating and pressurised irrigation, designed for frost control.

They have the necessary machinery and equipment to carry out field work efficiently (tractors, fumigators, dethatchers, shredders and other agricultural implements). This type of producer has the necessary guarantees to be subject to credit. Because they have good economic and moral solvency and good business management, they have guarantees and preferential rates.

The benefit-cost ratio of these producers is very high; they invest $1 peso and get $3. The cost of production per hectare is around $98,000. Yields average 60.0 tonnes or more. On average, the production cost of one kilogram of apples is $1.64 pesos. This type of grower, although few in number, is well suited to provide them with any type of credit.

5.2 Importance of apple (*Malus domestica*) production in Cuauhtemoc, Chihuahua in the national context.

Apple production in 2013 generated a national value of around $4,265.3 million pesos (equivalent to US$277.6 million).6 million), participating in the creation of this economic flow a total of 23 states, within which the state of Chihuahua stood out, as it was the main contributor, contributing with almost 76% of the production value of the fruit at the national level, and very specifically, the DDR of Cuauhtemoc, Chihuahua, generated by itself a little more than a quarter (25.96%) of the national monetary income produced by the apple crop, the second leading state was Durango, which produced $9.33 out of every $100 produced by the crop in the country, the third place corresponded to the state of Coahuila, with 6.48% of the national production value, although it should be mentioned that in terms of the proportion of the national surface destined to this crop, according to Table 2, Chihuahua produces 4/5 of the national physical volume of apples, and the remaining 20% is produced by the remaining 22 states, within which Durango and Coahuila together produce 12.3% (with 7.64 and 4.66% respectively), so that the share of the remaining states is merely marginal in the national production.

The relative importance of the Cuauhtemoc RDA in the national production of red apples is provided by its participation percentages (shown in Table 2) in the harvested area, production and national GVP, since, on its own, the Cuauhtemoc RDA is the largest producer of red apples in the country.

DDR Cuauhtemoc participates with 13.13%, 20.1% and 24.95% of the harvested area, physical production and value of the national apple production.

Table 2 : Relative importance of the state of Chihuahua and Cuauhtemoc, Chihuahua in the national apple production economy

State	Area harvested		Annual production		Gross Value of Production		
	ha	% compared to the national level	Ton	% compared to the national level	Millions of pesos	USD$ in millions	% compared to the national level
Aguascalientes	48,0	0,08	376,6	0,04	$ 3,84	$ 0,25	0,09
Baja California	4,0	0,01	39,0	0,00	$ 0,23	$ 0,02	0,01
Chiapas	1.015,5	1,72	3.514,6	0,41	$ 27,39	$ 1,78	0,64
Chihuahua	26.882,1	45,41	684.669,9	79,74	$ 3.238,48	$ 210,77	75,93
Cuauhtemoc	7.770,00	13,13	172.593,00	20,10	$ 1.106,77	$ 72,03	25,95
% with respect to Chihuahua	28,90		25,21		34,18	34,18	
Coahuila	7.018,00	11,85	39.969,64	4,66	$ 276,38	$ 17,99	6,48
Federal District	113,50	0,19	715,55	0,08	$ 7,88	$ 0,51	0,18
Durango	9.763,00	16,49	65.596,36	7,64	$ 397,78	$ 25,89	9,33
Guanajuato	21,00	0,04	42,00	0,00	$ 0,20	$ 0,01	0,00
Guerrero	40,50	0,07	156,48	0,02	$ 0,92	$ 0,06	0,02
Hidalgo	866,50	1,46	3.688,92	0,43	$ 31,56	$ 2,05	0,74
Jalisco	22,50	0,04	110,15	0,01	$ 1,48	$ 0,10	0,03
Michoacan	143,00	0,24	1.030,72	0,12	$ 5,60	$ 0,36	0,13
Morelos	6,00	0,01	72,00	0,01	$ 0,60	$ 0,04	0,01
Mexico	143,10	0,24	827,96	0,10	$ 4,99	$ 0,32	0,12
Nuevo Leon	1.197,60	2,02	3.852,74	0,45	$ 39,08	$ 2,54	0,92
Oaxaca	556,25	0,94	1.789,58	0,21	$ 8,40	$ 0,55	0,20
Puebla	8.571,98	14,48	35.857,30	4,18	$ 141,55	$ 9,21	3,32
Queretaro	575,50	0,97	1.065,60	0,12	$ 4,61	$ 0,30	0,11
San Luis Potosi	20,00	0,03	152,50	0,02	$ 1,14	$ 0,07	0,03
Sonora	160,69	0,27	1.307,38	0,15	$ 5,88	$ 0,38	0,14
Tlaxcala	14,00	0,02	119,50	0,01	$ 1,24	$ 0,08	0,03
Veracruz	824,50	1,39	6.857,96	0,80	$ 20,18	$ 1,31	0,47
Zacatecas	1.191,70	2,01	6.795,47	0,79	$ 45,88	$ 2,99	1,08
National	**59.198,9**	**100,00**	**858.607,9**	**100,00**	**4.265,3**	**277,6**	**100,00**

Source: Own elaboration, based on figures from Sistema de Información Agroalimentaria y Pesquera (SIAP).

Table 3 shows the prices and the physical and monetary yields per hectare for the 23 states, and the national concentrate, for the production of red apple, DDR Cuauhtemoc is marked in red.

Table 3: Position of Cuauhtemoc, Chihuahua in yield and price in the national context of red apple production.

State	Physical yield (ton ha)-1	Price (USD$) ton-	Monetary yield (USD$ ha)-1	National physical performance =1	Domestic price =1	Monetary income per ha national =1
Aguascalientes	7,85	$ 662,9	$ 5.201	0,54	2,05	1,11
Baja California	9,75	$ 388,5	$ 3.788	0,67	1,20	0,81
Chiapas	3,46	$ 507,1	$ 1.755	0,24	1,57	0,37
Chihuahua	25,47	$ 307,8	$ 7.841	1,76	0,95	1,67
Cuauhtemoc	22,21	$ 417,4	$ 9.271	1,53	1,29	1,98
% with respect to Chihuahua	87,2	135,6	118,2			
Coahuila	5,70	$ 450,0	$ 2.563	0,39	1,39	0,55
Federal District	6,30	$ 716,6	$ 4.518	0,43	2,22	0,96
Durango	6,72	$ 394,7	$ 2.652	0,46	1,22	0,57
Guanajuato	2,00	$ 312,4	$ 625	0,14	0,97	0,13
Guerrero	3,86	$ 383,1	$ 1.480	0,27	1,18	0,32
Hidalgo	4,26	$ 556,8	$ 2.370	0,29	1,72	0,51
Jalisco	4,90	$ 871,9	$ 4.268	0,34	2,70	0,91
Michoacan	7,21	$ 353,4	$ 2.547	0,50	1,09	0,54
Morelos	12,00	$ 544,0	$ 6.528	0,83	1,68	1,39
Mexico	5,79	$ 392,5	$ 2.271	0,40	1,21	0,48
Nuevo Leon	3,22	$ 660,2	$ 2.124	0,22	2,04	0,45
Oaxaca	3,22	$ 305,6	$ 983	0,22	0,95	0,21
Puebla	4,18	$ 256,9	$ 1.075	0,29	0,79	0,23
Queretaro	1,85	$ 281,5	$ 521	0,13	0,87	0,11
San Luis Potosi	7,63	$ 486,2	$ 3.708	0,53	1,50	0,79
Sonora	8,14	$ 292,9	$ 2.383	0,56	0,91	0,51
Tlaxcala	8,54	$ 672,8	$ 5.742	0,59	2,08	1,22
Veracruz	8,32	$ 191,5	$ 1.593	0,57	0,59	0,34
Zacatecas	5,70	$ 439,4	$ 2.506	0,39	1,36	0,53
National	14,50	$ 323,3	$ 4.689	1,00	1,00	1,00

Source: Own elaboration, based on 14 April 2015 at 1256 hours USD$. **Table 2. Exchange rate : $15.365 Mexican pesos per**

From this source, it is observed that the national averages for physical and monetary yields, as well as for prices at the national level, were of the order of 14.50 ton ha^{-1} , US$323.5 ton^{-1} in terms of price, and US$4,689 of monetary yield per ha, and when disaggregated for the state of Chihuahua it was found that with 25.47 ton ha-1 , it enjoyed the highest physical yield, but not the highest price, since with its US$417.4 ton^{-1} although it was above the national average price (US$323.3) was below the prices of apples produced by states such as Aguascalientes, Chiapas, Coahuila, Distrito Federal, Hidalgo, Jalisco, Morelos, Nuevo Leon, San Luis Potosi, Tlaxcala and Zacatecas, and although the price was not "good" in relation to other states, but above the national average, it managed to compensate the undesirable effect on income per hectare with its high physical yield, as the state of

Chihuahua, with US$7,841, had the highest monetary income per hectare, much higher than the other states, and 67% higher than the national average.

For its part, the DDR Cuauhtemoc, had a monetary income not only above the national average, but even above the state average, since the average ha of apples in Cuauhtemoc generated US$9,271 in income, almost double, 98% above, the national average, and this was achieved because Cuauhtemoc had a physical yield 53% higher than the national average, as well as a price 29% above the national average, although, it is insisted, in other states the price was higher (Table 3).

Of the remaining apple producers in the country, the states of Morelos and Tlaxcala stand out in terms of monetary income per hectare, with US$6,258 and US$5,742 respectively, although their contribution to national production is merely marginal, as mentioned above, but it should be noted that their high monetary yield is not due to their physical yield, which is very low in relation to that of Chihuahua or Cuauhtemoc, which are high, which is very low in relation to that of Chihuahua or Cuauhtemoc, which are high, but it is due to the high price, higher in both states than the average price of Chihuahua, i.e. in Chihuahua, the monetary productivity per ha is due to its physical productivity, while in Morelos and Tlaxcala it is due to the local price, higher than the national and Chihuahua prices (Table 3).

Table 4 takes up the figures of the main macroeconomic variables of apple production in Cuauhtemoc, such as harvested area, annual physical production, VBP, physical yield, prices and income per hectare, already analysed in previous paragraphs, but now these variables are linked to other interdependent variables such as prices and income per ha (which in turn are dependent precisely on area, production, value and price) as well as other independent variables such as costs and wages per ha and volumes of water used in production, production, value and price) as well as other independent variables such as costs and wages per ha and volumes of water used in production, and in this way, new economic variables such as profit per ha, employment generated, and above all, water and capital productivity can be obtained.

Table 4 shows a notorious difference between producers with low, medium and high use of technology in production, since some variables, such as physical yield, are characterised by the fact that while a hectare of apple, in a producer with low technology, yields 10 tons on average, while in the medium one generates 22.21 and in the high use of technology that same hectare of apple produces 35 tons, together with differentiated prices of US$280.6, US$382.6 and US$417.4 ton^{-1} , respectively for low, medium and high technology use producers, meant that while a low technology use producer generated an income of US$2,806 per ha, a medium producer produced US$8,498 and a high technology use producer produced US$14,607 ha^{-1} , that is, according to the index numbers on the right side of Table 4, a low technology use producer generates an income equivalent to only 33% of that produced by the medium producer, and a high technology use producer generates an income equivalent to only 33% of that produced by the medium producer, while a high technology use producer generates an income equivalent to the medium producer.

of that produced by the median, while the high-tech producer generated 72% more income than the average producer.

Table 4: Area, production, production value, income, costs, profitability, water used and employment in low, medium and high technology producers of apple (*Malus domestica*) in Cuauhtemoc, Chihuahua, Mexico in the 2013 agricultural cycle.

Macroeconomic variable	A) Low-tech producers	B) High-tech producers	C) Medium technology level producers	A/C	B/C
Harvested area (ha)	no data	no data	7.770,00		
Annual physical production (tonnes)	no data	no data	172.593,00		
VBP (Millions of USD)	no data	no data	$ 72,03		
Ton/ha	10,00	35	22,21	0,45	1,58
Price (USD) ton^{-1}	$ 280,6	417,4	382,6	0,73	1,09
Income (USD) ha^{-1}	$ 2.806	$ 14.607	$ 8.498	0,33	1,72
Cost (USD)ha^{-1}	$ 1.878	$ 7.851	$ 5.366	0,35	1,46
Cost (USD) kg^{-1}	$ 0,19	$ 0,22	$ 0,24	0,78	0,93
Profit (USD) ha^{-1}	$ 928	$ 6.757	$ 3.132	0,30	2,16
Gain (USD) kg^{-1}	$ 0,09	$ 0,19	$ 0,14	0,66	1,37
Benefit/Cost Ratio	1,494	1,861	1,584	0,94	1,17
# of daily wages/ha (direct)	89	72	62	1,44	1,16
Kg of apples per day	112,4	486,1	358,3	0,31	1,36
Monetary gain/day	$ 10,4	$ 93,8	$ 50,5	0,21	1,86
Net irrigation sheet (m)	1,10	1,10	1,10		
Volume of water used/ha (in m)^3	11.000	11.000	11.000		

Volume of water used over the whole harvested area (in hm3 =1 million m^3 = 1000 million L)	Sd	Sd	85,47		
Total monetary gain (Millions of USD$)	Sd	Sd	$ 24,34		
Total daily wages per year	Sd	Sd	481.740		
Number of permanent jobs/year (1 permanent job = 6 days/week for 48 weeks/year)	Sd	Sd	1.673		
Capital investment (Millions of USD$)	Sd	Sd	$ 41,69		
Source: Own elaboration, based on SIAP figures for harvested area, annual physical production and VBP, and Table 6.					

The generation of income per ha is, per se, important, but more important is the generation of profit, as Table 4 shows that the profit per ha^{-1} , was in the order of 928, 6,757 and US$3,132 respectively for low, high and medium technology use farmers. In relation to the average producer, it was determined that a low technology use producer was 70% below the average producer (the index on the right hand side of Table 4 was 0.30) while the high technology use producer was 116% above (the indicator was 2.36). The Benefit-Cost Ratio (BCR) shown in Table 4 indicates that in the three technological profiles of Cuauhtemoc apple production it was possible to recover the investment and also to obtain surpluses. Those with the lowest RB/C were the producers of low technological level, in which the indicator was equal to 1.494, the intermediate ones had a RB/C of 1.584, while those with high use of technology showed an indicator of 1.861, which indicates that they were able to recover each dollar invested in the production, and in addition, they achieved a profit of 49.4, 58.4 and 86.1 cents for each dollar invested.

In relation to the second apple producing state, the state of Durango, specifically the region of Canatlan, according to Barrientos (2015)[40] , in 2013, the indicators of the RB/C of the producers of low and high technological level in that location, were 1.76 versus 1.972, figures within which the high-tech producers of Cuauhtemoc are located, but which exceed the profitability of the low- and medium-tech producers, the cause of this lies in the costs of production, While a prototype producer of high technology use in

[40] Barrientos, D. L. Isabel, (2015). Productivity of water, soil, capital and labour in apple (*Malus domestica*) cultivation in Canatlan, Durango, Mexico. Bachelor's thesis. Universidad Autonoma Chapingo, Unidad Regional Universitaria de Zonas Aridas, Bermejillo, Durango, Mexico. See Table 5, page 30.

Cuauhtemoc, Chihuahua, had a cost per hectare in the order of US$7,851, those of high technology use in Canatlan, Durango, had a cost per hectare in the order of US$2,987.This is 2.63 times more expensive in Cuauhtemoc than in Canatlan, and this is because the "high" technology users in Canatlan actually tend to resemble the poor producers in Cuauhtemoc, i.e. the degree of technology is much higher in Chihuahua than in Durango.

In relation to the same apple crop, in the same state of Chihuahua, and similar levels of technology, but with profitability data for the period 2001-2002, Ramkez et al (2006)[41] , indicate that the profitability indices for producers with high, medium and low use of technology were 2.38, 1.86 and 1.37, while Table 4 of this study, for the year 2013, indicates that these profitability indicators were of the order of 1.861, 1.584 and 1.494 respectively, which would indicate that in the time span from 2001-2002 to 2013, while the producers of high and medium use of technology lost profitability, those of low technological use increased it.

The reason that high and medium technology-using farmers decreased their profitability between 2002 and 2013, while low technology-using farmers increased their profitability was due to three causes: the first two are physical yield per hectare and price (sources of income per hectare, which comes from multiplying physical yield by price), the third cause was production costs per hectare. Thus, Table 5 shows that among producers with high (HT) and medium (MT) technology use, while income per hectare was multiplied by 1.97 and 1.94 times, costs per hectare were multiplied by 2.52 and 2.29 times respectively. In other words, although both types of producers (AT and MT) increased their yields and prices, thereby increasing their income per ha, the respective increases in their costs, by a greater percentage than the percentage by which their income increased, caused them to lose profitability.

Table 5: Comparison of profitability indices per hectare between 2002 and 2013 of the apple crop in Cuauhtemoc, Chihuahua between producers with high (AT), medium (MT) and low (BT) use of technology. Monetary figures in nominal pesos.

[41] Ramirez-Legarreta, M. R., Jacobo-Cuellar, J. L., Avila-Marioni, M. R., & Parra-Quezada, R. A. (2006). Crop losses, production efficiency and profitability of apple orchards with varying degrees of technification in Chihuahua, Mexico. *Revista Fitotecnia Mexicana*, *29*(3), 215-222.

Technological profile	Variable	A)2002	B) 2013	B/A
AT	(a) kg per ha	32.584,0	35.000,0	1,07
	(b) Price per kg	$3,50	$6,41	1,83
	(c) Income per ha	$ 114.044	$ 224.467	1,97
	(d) Cost per ha	$ 47.964	$ 120.631	2,52
	profitability ratio=c/d	**2,38**	**1,86**	
MT	(a) kg per ha	19.236,0	22.210,0	1,15
	(b) Price per kg	$ 3,50	$5,88	1,68
	(c) Income per ha	$ 67.326	$ 130.565	1,94
	(d) Cost per ha	$ 35.949	$ 82.400	2,29
	profitability ratio=c/d	**1,86**	**1,58**	
BT	(a) kg per ha	9.440,0	10.000,0	1,06
	(b) Price per kg	$ 3,50	$4,31	1,23
	(c) Income per ha	$ 33.040	$ 43.114	1,30
	(d) Cost per ha	$ 23.486	$ 28.855	1,23
	profitability ratio=c/d	**1,37**	**1,49**	

Source: Own elaboration, based on Table 5 of Ramfrez *et al* (2006) for the year 2002, and conversion to Mexican pesos of the cost per hectare recorded in our Table 4, considering the exchange rate of $15.365 Mexican pesos per USD as of 14 April 2015 for the interbank dollar of the Bank of Mexico.

The reason for this is actually simple: the increase in the cost of energy on which these two types of production are heavily dependent for pumping, but not the LV producers, among whom the income per hectare increased by 30% (the indicator in Table 5 is 1.30) while the cost increased by only 23%, the cause is the same as for LV and MV producers, only that it worked in the opposite direction for them, since, by not depending so much on electricity for pumping, as they use rainwater, it had the effect of increasing their profitability.

The social productivity of labour, in physical terms, was characterised by the fact that the same working day yielded different amounts of product, so that in poor producers, one working day represented 112.4 kg of apples, while the average producer produced in the same time 358.3 kg, while the producer with a broad technological base produced 486.1 kg. This suggests that a highly technified producer produces 36% more apples than the average producer, while the low technified producer produces less than a third (31%) more apples than the medium technified producer (Table 4). With the same irrigation sheet, 1.1 m, the water demand per ha was similar for all three types of farmers: 11,000 m^3 (Table 4).

Table 6 records the costs per ha for each of the three technical production profiles, i.e. low, medium and high technology producers used in apple production.

Table 6: Production costs per ha in the cultivation of red apple (*Malus domestica*) in Cuauhtemoc, Chihuahua in producers with low, medium and high use of technology.

COST ITEM	Technological level						% in relation to total cost		
	Cost per ha in MX$			Cost per ha in US$					
	Under	High	Medium	Under	High	Medium	Under	High	Medium
Ground tracking and skirting	1.400,0			91,1			4,9	0	0
Winter pruning (trees)	3.000,0	4.000,0	4.000,0	195,2	260,3	260,3	10,4	3,3	4,9
Pruning clean-up (daily wages)	300,0	280,0	280,0	19,5	18,2	18,2	1,0	0,2	0,3
Soil and foliar analysis	400,0	45,0	45,0	26,0	2,9	2,9	1,4	0,0	0,1
Winter Oil (Lts)	600,0	858,0	858,0	39,0	55,8	55,8	2,1	0,7	1,0
Cold compensator (Litres)		1.794,0	1.794,0		116,8	116,8	0,0	1,5	2,2
Foliar fertilisation (K,Zn,B)	350,0	5.206,6	4.788,9	22,8	338,9	311,7	1,2	4,3	5,8
Soil Fertilisation (N,P,K,Zn,B,Fe,Mg)	1.500,0	7.223,1	5.778,5	97,6	470,1	376,1	5,2	6,0	7,0
Post-harvest foliar fertilisation (N)	150,0	198,3	198,3	9,8	12,9	12,9	0,5	0,2	0,2
Pollination (Hive rental)		1.290,0	860,0		84,0	56,0	0,0	1,1	1,0
Knob control (with appl.)	1.590,0	3.864,9	3.864,9	103,5	251,5	251,5	5,5	3,2	4,7
Mite control (Agrimec)		3.992,7	3.992,7		259,9	259,9	0,0	3,3	4,8
Release of beneficial insects		220,0	220,0		14,3	14,3	0,0	0,2	0,3
Rosette Control (Russeting)		1.612,0	2.418,0		104,9	157,4	0,0	1,3	2,9
Control of fire blight (Erwinia)	600,0	2.022,7	1.011,3	39,0	131,6	65,8	2,1	1,7	1,2
Control of powdery mildew (powdery mildew)	1.000,0	3.345,5	1.672,8	65,1	217,7	108,9	3,5	2,8	2,0
Thinning / chemical thinning		1.712,5	1.052,5		111,5	68,5	0,0	1,4	1,3
Manual thinning (trees)	2.000,0	4.000,0	4.000,0	130,2	260,3	260,3	6,9	3,3	4,9
Calcium fruit firmness	394,0	1.576,0	1.576,0	25,6	102,6	102,6	1,4	1,3	1,9
Calcium applications	250,0	1.000,0	800,0	16,3	65,1	52,1	0,9	0,8	1,0
Chemical weed control		500,0	500,0		32,5	32,5	0,0	0,4	0,6
Mechanical weed control	375,0	400,0	300,0	24,4	26,0	19,5	1,3	0,3	0,4
Red dot control		2.194,7	2.194,7		142,8	142,8	0,0	1,8	2,7
Pre-harvest drop control / fixative		689,0	689,0		44,8	44,8	0,0	0,6	0,8
Fumigation work (daily wages)	1.500,0	2.500,0	2.000,0	97,6	162,7	130,2	5,2	2,1	2,4
Irrigation/ well energy (month)		3.600,0	3.600,0		234,3	234,3	0,0	3,0	4,4
Collection (tonne)	3.500,0	12.250,0	8.750,0	227,8	797,3	569,5	12,1	10,2	10,6
Transport to packer (ton.)	750,0	3.500,0	2.500,0	48,8	227,8	162,7	2,6	2,9	3,0
Frost control defence (lts. Diesel)		33.000,0	7.200,0		2.147,7	468,6	0,0	27,4	8,7
Frost protection (daily wages)		3.750,0	1.800,0		244,1	117,1	0,0	3,1	2,2
Laying and removal of netting (daily wages)		3.000,0	2.700,0	-	195,2	175,7	0,0	2,5	3,3
Hail netting work	1.800,0			117,1			6,2	0,0	0,0
Irrigation - pumping / sprinkler	2.400,0			156,2			8,3	0,0	0,0
Holidays, Christmas bonuses and gratuities		2.500,0	2.500,0		162,7	162,7	0,0	2,1	3,0
Labour taxes, medical services, etc.		2.500,0	2.500,0		162,7	162,7	0,0	2,1	3,0
Technical advice (annual)		1.000,0	1.000,0		65,1	65,1	0,0	0,8	1,2
SUBTOTAL	$23.859	$115.625	$77.444	$1.553	$ 7.525	$5.040	82,7	95,9	93,9
Agricultural insurance	5000	5000	5000	325,4	325,4	325,4	17,3	4,1	6,1
TOTAL	$28.859	$120.625	$82.444	$1.878	$7.851	$5.366	100,0	100,0	100,0
Day rates per ha	89	72	62				0,0	0,0	0,0

Source: Own elaboration, based on the production costs per ha reported by UNIFRUT of Cuauhtemoc, Chihuahua, to which the cost of agricultural insurance was added, which UNIFRUT does not take into account since it indicates that these correspond to an expense that the farmer will have to make on his own.

That according to the source of the costs, the company UNIFRUT of Chihuahua, Chihuahua, they were of the order of $23,859 (equivalent to US$1,553), $77,444 (equivalent to US$5,040) and $115,625 (equivalent to US$7,525) respectively for the producers of low, medium and high use of technology, nevertheless, it was decided to add $5,000 (equivalent to US$325.4) to each, as the same cost sheet provided by UNFRUT indicates that $5,000 of agricultural insurance should not be added, as this amount should be covered by the farmer, so, in the end, this amount should be added.

Ash, the costs per hectare rise, for each of the three types of producers mentioned above, in the same order as above, by $28,859 (US$1,878), $82,444 (US$5,366) and $120,625 (US$7,851). It should be noted among the different cost component items, that while in the high and medium technology use items the "Irrigation/well energy" item does have an amount, the low technology use ones do not, suggesting that in reality they depend on both gravity irrigation water from local dams and rainfall to carry out production, but not the medium and high technology use ones, where pumped irrigation is essential.

The right hand side of Table 6 shows the most important items for each production profile, so it was found that in a low-income producer, the top five cost components were agricultural insurance 17.3%, harvesting 12.1%, winter pruning 10.4%, irrigation cost - pumping/watering and manual tree thinning 6.9%. In high-tech producers, the main components of their production cost were frost protection with 27.4% of the total cost, harvesting with 10.2%, soil fertilisation with 6.0% and foliar with 4.3%, and finally agricultural insurance with 4.1%.1%, and in the producers of medium technological use, were the same items as in those of high use: defence against frost with 8.7% of the total cost, harvesting with 10.6%, soil fertilisation with 7.0% and foliar with 5.8%, and finally agricultural insurance with 6.1%. The only items of the total cost in which the three technology levels coincided in relative importance were crop insurance, although, not equally important for each, it represented almost a fifth of the cost for the poor farmer, while for the "rich" farmer only 4.1%, the same as the harvesting item, but, like the crop insurance component, it was a higher percentage for the low-use farmers.

(12.1%) and 10.2% and 10.6% in high and medium (see Table 5).

5.3 Indicators of soil, water, capital and labour productivity in the production of red apple (*Malus domestica*) in Cuauhtemoc, Chihuahua in producers with low, medium and high use of technology.

Soil productivity among the three types of producers ranges between 10 and 35 ton ha^{-1} , this is due to the characteristics of each type of producer, as stated in section 5.1 stated, also, the fact that they do not have access to technology, do not have enough organization and lack of adequate administrative organization, makes that the producers with low use of technology generate the lowest income and profit per hectare (US$2,806 and US$928 respectively), as shown in Table 7, but not the producers with high use of technology, who due to their characteristics are those who generate the highest income and profit per hectare: US$14,607 and US$6,757 respectively, income and profit per hectare.

The intermediate technology level producers, in terms of yield, income and profit per hectare, are precisely between those of low and high technological use, and are about 55% of the apple producers, as stated in section 5.1, they had a yield of 22.21 ton ha^{-1} , and generated income of the order of US$8,497 ha^{-1} , of which US$3,132 was the profit per ha (see Table 7).

Table 7: Indicators of the productivity of water, labour and capital in the production of the red apple crop in Cuauhtemoc, Chihuahua in 2013. Monetary figures in USD on 14 April 2015 at 1256 hours, at the rate of $15.365 Mexican pesos per US dollar.

Economic variable		A) Producers at level low technological	B) Producers at level high technological	C) Producers at level medium technological	A/C	B/C
Soil productivity:						
Physical performance	ton ha^{-1}	10,00	35,00	22,21	0,45	1,58
Monetary return (USD$)	Income ha^{-1}	$ 2.806	$14.607	8497,71	0,33	1,72
Monetary return (USD$)	Profit ha^{-1}	$ 928	$6.757	$ 3.132	0,30	2,16
Water productivity:						
physical performance	kg $_{m3}$	0,91	3,18	2,02	0,45	1,58
physical performance	L kg^{-1}	1.100	314	495	2,22	0,63
monetary performance	US$ profit hm 3^{-}	$ 84.341	$ 614.244	$ 284.726	0,30	2,16
Social productivity of water	Employment hm^{-3}	28,1	22,7	19,6	1,44	1,16
Capital productivity:						
RB/C		1,494	1,861	1,584	0,94	1,17
Rate of return	% of surplus on invested capital	49,4%	86,1%	58,4%	0,85	1,47
Social productivity of capital (Jobs generated per million dollars invested)	Employment/Investment	164,5	31,8	40,1	4,10	0,79
Break-even point "EP	ton ha^{-1}	6,69	18,81	14,03	0,48	1,34
Physical performance/PE	dimensionless	1,49	1,86	1,58	0,94	1,17

Labour productivity:						
Work per ha	Days/ha	89	72	62	1,44	1,16
Work per ha	Hours/ha	712,0	576,0	496,0	1,44	1,16
Working hours per tonne	h ton^{-1}	71,20	16,46	22,33	3,19	0,74
Kilograms per hour	kg h^{-1}	14,0	60,8	44,8	0,31	1,36
Earnings per day	USD$ per day^{-1}	$ 10,4	$93 ,8	$50,5	0,21	1,86
Hourly earnings	USD$ hour^{-1}	$ 1,30	$11,73	$6,31	0,21	1,86
Source: Own elaboration, based on figures in Tables 4 and 6.						

The indicators of water productivity in apple cultivation in Table 7 show that the same volume of water, one cubic metre, produced different amounts of apple in each of the three types of producers, as it was found that while in a producer with low technological use, that volume of water generated 0.91 kg of apple, while in the intermediate producer it produced 2.02 kg and, finally, in the producer with high technological endowment it was 3.18 kg.

Therefore, taking the intermediate producer as a benchmark, it was observed that a producer with low technology use produced only 45% of what the medium producer generates of apples with that cubic metre, but, in contrast, the one with high technology use produced 2.02 times, that is, 102% more apples per cubic metre of water than the regional average.

The above immediately raises the question: ^Why is it that there were such different physical yields per cubic metre of irrigated water? The answer to this question has to do, in the first place, with the two variables on which this productivity variable depends, which measures the amount of kg of apples produced per cubic metre of irrigated water, that is, the physical yield per hectare and the volume of irrigated water per hectare. The first variable, the physical yield per hectare depends, in turn, on the degree of production technology, and as it was recorded in section 5.1, when using on a very small scale the producers of low technology use, their yield is very "small" in relation to those producers who use more technology: 10, 22.21 and 35 ton ha^{-1} . It should be remembered that the low-tech producers (about one third of the total producers) have neither high economic resources nor technical assistance, are usually not subject to credit, lack the adequate tools to implement a technological package, do not have hail prevention systems, This means that their yields range from 6 to 15 tons per hectare maximum, but not the large producers, around 14%, who have protection against frost and hail, have credit, technical assistance and high availability of economic resources, achieve average yields of 30 to 60 tons per hectare^{-1} and sometimes more. The second variable on which the amount of kg of apples produced per cubic metre of irrigated water depends, is the volume of water used per ha, which although it was assumed to be the same, given that the crop has the same water requirement, and although the same volume of water is used in producers with low, medium or high use of irrigation technology, It is not the same that the cultivar receives the water every few days, or even worse, that it depends on rainwater, than to manage that volume "V" of water that the cultivar demands, by providing it with water in the most critical stages for the apple tree.

The water productivity determined in this study, which uses historical production data, differs and at the same time does not differ from the water productivity figures for apple cultivation in the same region, based on experimental data to see the effect of controlled irrigation deficit, it differs in that while in this study it was determined that one cubic metre of water produces on average 3.18, 2.12 and 0.91 kg m^{-3} for producers with high, medium and low use of technology in production, Quezada *et al,* (2009)[42] found that for apple trees with controlled irrigation deficit, the productivity in water use triples, from 2.39 to 7.68 kg m^{-3} , but on the other hand, the aforementioned author, when considering 2.39 kg m^{-3} as a reference parameter against which the experimental apple tree with controlled irrigation deficit is compared, approaches the 2.02 kg m^{-3} emanated from this study for the average producers, although, the high technology producers of our work, with 3.18 kg m, would have a productivity only 50% of that obtained experimentally.

Comparing now the results found for Cuauhtemoc, Chihuahua, of 0.91, 2.12 and 3.18 kg m^{-3} of apple, for low, intermediate and high technology producers respectively, shown in table 7, in relation to the world average of water productivity in apple (and other products), as reported by Chapagaine and Hoekstra (2004)[43] (from

[42] Quezada, R. A. P., Franco, P. O., Alvarez, J. P. A., & Sanchez, N. C. (2009). Apple tree productivity and growth under controlled deficit irrigation. *Terra Latinoamericana*, *27*(4), 337-343.
[43] Chapagain, A. K., & Hoekstra, A. Y. (2004). Water footprints of nations, value of water research Report Series No. 16.

whom their figures were taken as the basis for Table 8 of this study) who report that 70 litres are required to produce a 100g apple, i.e., converted to kg m^{-3} , the water productivity, in that world average, is 1.829 kg m^{-3} , so it is easy to see that the water productivity of the producers of low technological use of this study showed only 49.7% (=0.91/1.829=0.497) of the world average productivity level, but, both the intermediate and medium technology use producers of Cuauhtemoc, were above the world average, since the medium use producers were 15.9% (=2.12/1.829=1.159) more productive in water use than the world average, and those with high technological endowment were 73.9% (=3.18/1.829=1.739) above the world average of productivity in water use in apple cultivation.

Table 8: Global average water footprint of selected commodities

Product	Unit	Virtual water content (litres)
1 slice of bread	30 g	40
1 potato	100 gr	25
1 apple	**100 gr**	**70**
1 tomato	70 gr	13
1 egg	40 gr	135
1 hamburger	150 gr	2400
1 glass of beer	250 ml	75
1 glass of milk	200 ml	200
1 cup of coffee	125 ml	140
1 glass of wine	125 ml	120
1 glass of orange juice	200 ml	170
1 pair of shoes	leather	8000
1 cotton t-shirt	medium, 500 gr	4100
1 sheet of A4 paper	$80 g/m^2$	10
1 microchip	2g	32

Source: Own elaboration, based on information from: Chapagaine, A. K. and Hoekstra, A. Y. (2004) "Water footprints of nations", Value of Water Research Report Series No. 16, UNESCO-IHE, Delft, the Netherlands. at:http://waterfootprint.org

The inverse variable to the one discussed in the two preceding paragraphs, which is the variable that measures the amount of water, in litres, necessary to produce one kg of product, which would normally result in what is called the water footprint, is shown in Table 7, showing that the water footprint of the apple cultivar produced in low technology with 1,100 litres kg^{-1} , was 2.22 times the water footprint of the apple produced by the average producer, with 495 litres kg^{-1} , but, in short, the high degree of technology significantly reduced the water footprint of the crop, since the apple produced with a high degree of technology implied the consumption of only 314 litres to produce the same kg of apple that, it is insisted, required 1,100 litres at low technology.

In the economic aspect, the social productivity of irrigation water was characterized, according to Table 7, in that by using one hm-3 (remember that one cubic hectometre has one million cubic metres), it was possible to generate a gross profit of US$84,341 in the production of apples with low technological use, equivalent to only 30% of what that volume of water generated a profit for the average grower, at US$284,726, but still far short of the US$614,244 that the same volume of water generated a profit for the elite growers in terms of technology.

In the field of agricultural economics, the important part is the economy, since it provides the instruments that will help the analysis, while agriculture helps by providing only the raw material on which the analytical-economic instruments will be based. Although they do not refer to apple tree cultivation, the figures of Aldaya, Novo and Lamas (2011[44]) allude to the value or monetary income produced per unit volume of water, the economic productivity of water, which they call apparent water productivity, in different crops such as olive trees, vines, cereals and others, and point out that one m-3 of water has an apparent productivity ranging from €0.1 in cereals to €5.9 in strawberries, with olive trees in between, with €0.2 and vines with €0.3, so, when converting to USD hm^{-3} , these figures of the authors Aldaya, *et al,* (2011), to compare with the figures found in our study, it was found that cereals, strawberries, olives and vines, respectively, have an apparent productivity of US$112,540, US$6, 639,869, US$225,080 and US$337,620 respectively, which when compared against the Cuauhtemoc apple tree figures (see Table 7) of US$284,726 and US$614,244 for medium and high technology producers, would fall within the range of the figures of Aldaya et al op cit, but it should be

[44] Aldaya, M. M., Novo, F., & Llamas, M. (2010). Incorporating the water footprint and environmental water requirements into policy: reflections from the Donana Region (Spain). *Re-thinking Water and Food Security, CRC Press/Balkema, London,* 193-218.

noted that in the case of this study, The monetary figures refer to the profit per cubic hectometre, and not to the income or value generated per cubic hectometre as in the case of the authors mentioned above, so that our results are validated by comparing them with the olive tree and vine described by the authors, and furthermore, by considering on our part only the income (which is always greater than the profit when it is positive), would tend to bring the figures of US$ profit per cubic hectometre in Table 7 closer to those of Aldaya *et al.* for strawberries*.,* (2010).

Having clarified that the figures in table 7 are USD profit per hm, and those in aldaya et al are income per hm^3 , it then remains to convert the Cuauhtemoc and Aldaya figures to a single unit of measurement: US$ income per cubic metre. Ash in the case of Cuauhtemoc apple trees, one cubic metre generated an income of US$0.772 per cubic metre for medium technology users, while for high technology users this volume of water generated $1.327, while Aldaya *et al*'s (2010) figures ranged from US$0.11254 (cereals), US$6.639 (strawberry), US$0.225 (olive) and US$0.337 in grapevine, which validates the apparent water productivity figures we found for the Cuauhtemoc apple tree, by positioning its indicators in the middle of the Aldaya *et al* (2010) figures, or better still, closer to those of strawberry than those of cereals.

While there is no worldwide data on the economic productivity of irrigation water in terms of the amount of profit produced per cubic hectometre of irrigated water in apple, there is such data for apple cultivation, although referring to the state of Durango, Mexico, the second most important state, after Chihuahua, in terms of contribution to the national production of this crop. In 2013, low and high technology apple producers in that state registered profits of US$148,405 and US$362,825 per cubic hectometre, according to Barrientos (2015)[45] , which in relation to our Table 7, which shows US$84,341 and US$614,244 respectively for that type of producers, allows inferring that a low-income producer in Cuauhtemoc, using the same volume of water, one cubic hectometre, produces only 56.8% the amount of profit than the same type of producer in the

[45] Barrientos, D. L. Isabel, (2015). *op cit*, see Box 5, page 30.

state of Durango, but, in the case of highly technified producers, the average producer with that level of technology, in Chihuahua, when using that same volume of water in irrigation, one cubic hectometre, generates 69.3% more profit than the most technified producer in Durango.

In relation to the physical productivity of irrigation water, in the same apple crop, but in another Mexican state, Durango, Barrientos (2015[46]), determined that among low and high technology producers, when using the same volume of water in irrigation, one cubic meter, 0.88 and 1.88 kg of apple, while in Cuauhtemoc, as shown in Table 7, for the same type of producers, water productivity was 0.91 and 3.18 kg m^{-3} , from which it can be inferred that while among the producers of low and high technological level of both states, the low level of Cuauhtemoc is 3.4% more productive than that of Durango and the high technological endowment is 69.1% more productive in Chihuahua.

The above suggests, firstly, that although the low technology level producers of Cuauhtemoc, Chihuahua, are less productive in physical terms in the use of irrigation water, they are more efficient, more productive in the use of water, in economic terms, but not the high technology level producers, since, both in physical and economic terms, they were more productive in the use of irrigation water in relation to their counterparts in the state of Durango. Secondly, because Chihuahua is the main apple producer in Mexico, and is specialized in the export market, especially in the case of producers with high use of technology, the high level of profit, and therefore, the higher productivity in the economic use of irrigation water, but not in the case of producers of low technological profile, which are dedicated to serve the market for juices, preserves and porridges, which demand lower quality fruit, thus having a lower price per ton, even than their counterparts in the state of Durango, since, among the producers of low technological level, it was concluded that in Durango, according to Barrientos (2015, op cit, see Table 5) the prices per ton were of the order of US$393 for both producers, while in Cuauhtemoc, the price differed

[46] Barrientos, D. L. Isabel (2015), *op cit*, see Box 5, page 30.

in US$280.6 and US$417.4 per tonne (see Table 7 of this study).

The social productivity of irrigation water, contrary to what happened with water productivity in physical and economic terms (litres per kg, kg per cubic metre and profit per hm-3), turned out to be higher in the "poor" producers, those who use little technology, since in that segment of producers, one hm-3 of irrigated water generated 28.1 permanent jobs, while, according to Table 7, in those with medium technology use, 19.6 were produced and in those with high technology use, 22.7 permanent jobs were produced.This, according to the indicators on the right-hand side of Table 7, suggests that the social productivity of irrigation water, in relation to the average producer, was 44% higher in low-technology-use producers, while in high-technology-use producers this productivity was only 16% higher than the average.

This was completely foreseeable, since precisely because they use little technology, low-technology producers favour live labour, i.e. the intensive use of labour, sometimes the only resource available to the poor producer, although, of course, at harvest time the family's work is not enough and they have to hire temporary labour. ECLAC (1982)[47] . states that while in peasant production - the low-tech producer in this case - the commitment of the head of the enterprise to the labour force is total, insofar as all or a good part of the labour used in production is of a family nature and only marginally incurs the hiring of labour, such as at harvest time, The same does not happen with the entrepreneur - the producer with high use of technology in this case - whose only link with the labour force is of a legal nature, that is to say, when using more technology, family labour is dispensed with in the first place, and secondly, production is mechanised precisely to get rid of labour, thus reducing the amount of labour in the producers with medium and high use of technology.

Up to this point it can be concluded that in physical and economic terms, apple production is superior in high-tech production, as it reduces the water footprint while increasing the amount of product per cubic metre of water, but this is not so^ in terms of social

[47] ECLAC. 1982. Economia Campesina y Agricultura Empresarial (Tipologia de Productores del Agro Mexicano). Mexico, Siglo XXI Editores

productivity, as socially it is more profitable to use water in low-tech apple production, as it generates employment.
The productivity of capital, according to Table 7, was undoubtedly higher in the production of high technology use, where the Benefit Cost Ratio was higher, with an index of 1.881, while in the medium producers the indicator was equal to 1.584, and finally, in the producers of low technology use, the profit rate was 49.4%. The above suggests that in the three profiles of producers the investment is recovered, but the profit rate on the invested capital is lower in the poor producers where 49.4% of the invested capital is earned, while in the producers with a high use of technology the profit rate was of the order of 86.1%, that is, a little more than 36 percentage units above, while the average of the producers is around a profit of 58.4%.
A very important economic variable is the capacity of the invested capital to generate jobs, as Table 7 shows that the same amount of invested capital, one million US dollars, generated 164.5 jobs in apple production at low technological level versus 31.8 jobs in apple production with high use of technology, and finally corresponded to an average of 40.1 jobs for each million US dollars invested in those producers with average technological endowment.
It is necessary to stress that one should not over-weight one single variable and disconnect it from the others, i.e. one should not isolate social productivity from capital and over-value it in such a way as to recommend apple production using low technology and to generate government policies that discourage apple production using high technology because the former generates a lot of employment and the latter does not. For example, if the main objective is to save water resources in order to use part of it for the future, then the high-tech apple would be the one to encourage, knowing that this would lead to higher unemployment, but if employment is essential to reduce crime rates, then small-scale production is the ideal one to encourage.
The break-even point in economics is an important issue for the producer, since it suggests the physical quantity to be produced in order to start making profits, so that, if the producer produces less

than that quantity, he will incur losses, since at the break-even point the revenue is equal to the cost. Thus, Table 7 shows that in the case of average producers in terms of technology use, the break-even point was 14.03 tons, but these producers had physical yields in the order of 22.21 tons, according to the upper part of the same table, which suggests that average producers had a physical yield 58% (the indicator was 1.58) above the break-even point, in contrast to producers with low technology use, in which the break-even point was 6.69 tons ha, but their physical yield was 1.58 tons ha.69 ton ha, but their physical yield was only 49% (the indicator was 1.49), which although indicates that they produced a profit, in relation to the regional average production leaves them in a bad position, since they are closer to the loss zone than the average producer, which makes it difficult for them to have access to credit, since the financial institutions are the first thing they look at: how much they produce per hectare and how much is the minimum that must be produced in order to make a profit. It is necessary to remember that in section 5.1 some of the characteristics of each of the producers were pointed out, and among those with low technological use was not having access to credit, now we know that it is precisely because they have a low break-even point.

The same does not happen with the producers with a high use of technology, since, while the break-even point for them was 18.81 ton ha^{-1} , their physical yield was 86 percentage units above this economic minimum, which allows them to continue in this virtuous circle of having access to credit because they have high production and have high production due to their access to credit lines, which was one of their characteristics pointed out in section 5.1 (Table 7).

Labour productivity, as well as soil, water and capital productivity, was different for the three levels of production technology. One of the main differences was found in the amount of labour hours invested per tonne, since while the average producer required 22.33 labour hours to produce one tonne of apples, the poor, low-technology producer required 3.19 times that amount of labour, requiring 71.20 labour hours, while the high-technology producer required only 16.46 labour hours per tonne.

A particular form of social productivity is that which relates to the investment in labour (social or average) per unit of physical product, tons in this case, in fact, this particular form of productivity is one of the bases on which the theory of international trade put forward by the classical economist David Ricardo is based, who gives greater precision to Adam Smith's analysis by demonstrating that mutually beneficial trade between countries is possible even when only comparative advantages exist, coming to the conclusion that absolute comparative advantages are a special case of a more general principle which is precisely that of comparative advantages. David Ricardo shows that from the notion of comparative cost (in hours of labour invested per unit of product) one can define patterns of specialisation, taking into account two elements: labour costs and the terms of trade between countries[48] .

The United States, due to its geographical location and climate, is a favourable place for apple production, which gives it comparative advantages in the production of this fruit, as it is known that "American exports are directed towards Canada, EU-15,

Japan and Hong Kong, comprising 75% of them diversified into lettuce, onions, tomatoes, broccoli, apples, oranges and grapes. However, the US is a net importer of fruits and vegetables due to increased domestic consumption of fresh vegetables. The United States Department of Agricultural reports, for 1997, a per capita consumption of 72 kg of fresh fruits and 79 kg if consumption is included.

(COOK, 1998)[49] , which indicates that the Chihuahua apple has a window of opportunity, due to its geographical proximity to the United States of America, since, as the domestic market in that country specialises in the production of vegetables, it requires fresh fruit, i.e., taking up the point made in the previous paragraph, although in the United States the amount of labour hours invested per tonne is lower, compared to 16.46 to 22.33 labour hours per tonne (high and medium technology producers respectively, see Table 7), Chihuahua has a comparative advantage given the

---

[48] Pontificia Universidad Catolica del Ecuador. Economia y Finanzas. http://www.puce.edu. ec/econom^a/efi

[49]Cook, R. L. (1998, June). International trends in the fresh fruit and vegetable sector. In *WCHR-World Conference on Horticultural Research 495* (pp. 143-156).

specialisation of US production in vegetables, however, it would be better for apple production in Chihuahua if this labour investment could be reduced, in order to cement the advantage already present.

A variant of the social productivity described and analysed in the preceding paragraphs is the one shown in table 7 for the amount of kg produced per working hour. Thus, it was determined that in the average producers, those with medium use of technology, one hour of live labour produced 44.8 kg of apples, while in the other two producer profiles a total of 14.0 and 60.8 kg were generated in the same working time, one hour. This suggests that in relation to the average producer, the one with low technology use has a productivity per hour of work of only 31% (the indicator in Table 7 was equal to 0.31) of the regional average productivity, while the producer with high technology use was 36% of the regional average (the indicator was 1.36).

In relation to the comparative advantages analysed, it shows that the production of apples by poor producers, those with low technological use, is in a disadvantageous situation to approach a possible international marketing structure, so their strategy has been to attend the domestic market with lower quality apples at lower prices (see Table 4, where the difference between the prices of each profile of producers is clearly shown: \$280.6, \$382.6 and \$417.4 USD per tonne), aimed at undemanding market strata, so that a large part of its production supplies the domestic juice production and to a lesser extent fresh consumption.

The different technological endowment had repercussions, as has been demonstrated, not only in gradients of physical productivity, as well as of labour investment per tonne, but also in the fact that there was a difference between producers in the different amount of gross profit produced per unit of labour invested. Thus, Table 7 shows that while in a poor producer, a day's work represented US\$10.4 profit, in the regional average this amount rose to US\$50.5, but, in the case of the highly technified producers, the profit rose to US\$93.8 per day's work.

VI. CONCLUSIONS AND RECOMMENDATIONS

6.1 Conclusions

The objective of determining indicators to visualise and compare the productivity of water, labour and capital in each of the three levels of technology in the production of red apple in Cuauhtemoc, Chihuahua, was achieved.

Based on the mathematical-economical models used and on the results obtained by this methodology, the first hypothesis is rejected, because although the producers of high technology use "AT" showed the smallest metric footprint 314 litres kg^{-1} , versus 495 and 1,100 litres kg^{-1} , in the producers MT (medium technology use) and BT (low technology use) respectively, as well as having the highest water productivity index 3.18 versus 2.02 (MT) and 0.91 (B) kg m^{-3} , were not in which more employment is generated by the use of a cubic hectometre of water in irrigation, corresponding to indicators of 28.1 jobs hm^{-3} for BT producers and 19.6 jobs hm^{-3} in MT producers and 22.7 jobs hm^{-3} in AT producers.

Based on the mathematical-economical models used and on the results obtained by this methodology, the second hypothesis is rejected, because although the producers with high use of technology "AT" showed the highest Benefit-Cost Ratio "RB/C", 1.861, versus 1.584 and 1.494 in the MT (medium use of technology) and BT (low use of technology) producers respectively, they were not the ones in which more employment is generated by the investment of one million USD in production, corresponding to indicators of 164.5 jobs per million USD for the BT producers and 40.1 jobs per million USD in the MT producers and 31.8 jobs per million USD in the AT producers.

Based on the mathematical-economical models used and on the results obtained by this methodology, the third hypothesis is not rejected, since the producers with high use of technology "AT" showed to have the highest amount of kg of apples produced per hour of work, 60.8, versus 44.8 and 14.0 in the MT (medium use of technology) and BT (low use of technology) producers respectively, also, the AT producers were the ones that produced more profit per hour of work invested in the production, corresponding to them indicators of US$11.73 of profit for each hour of work, in comparison with the MT and BT producers, that in the same lapse of time, one hour of work, produced profits of the order of US$6.31 and US$1.30 respectively.

It was found that in the period 2002-2013, profitability was decreasing for AT and MT producers, while for BT producers profitability was increasing, as the BR/C decreased from 2.38 to 1.86 for high technology use producers and decreased from 1.86 to 1.58 for medium technology use producers, while for low technology use producers the BR/C rose from 1.37 to 1.49. There are two immediate causes for this, as the BR/C depends on two independent variables: income and cost. In the period 2002-2013, income per hectare multiplied by 1.97, 1.94 and 1.30 among AT, MT and BT producers respectively, while cost per hectare multiplied by 2.52, 2.29 and 1.23 respectively, so, as the speed of growth in cost was higher than in income in AT and MT producers, they lost percentage units of profitability, not so in BT producers, where income increased at a faster rate than costs, which made them gain eleven percentage units in their RB/C. At the same time, it should be clear that income depends on two independent variables: the physical yield "RF" per hectare and prices "p", and although physical yields increased faster among TA and MT producers than among BT producers (RF increased by 1.07, 1.15 and 1.06 respectively), prices "p" increased in greater proportion among TA and MT than among BT producers (p multiplied by 1.83 in TA, by 1.68 in MT and by 1.23 in BT), but, at the same time, prices "p" increased in greater proportion among TA and MT producers than among BT producers (p multiplied by 1.83 in TA, by 1.68 in MT and by 1.23 in BT), but, as already noted, the faster increase in costs among high- and medium-technology-using producers relative to low-technology-using producers cancelled out the good effect of the advances in agricultural productivity coming from higher technology use (as measured by the advance in physical yields), and moreover, it also cancelled out the favourable evolution of the selling prices of their apples, as a result of their greater organisation and bargaining power, which seems to be leading to a vicious circle between medium and high technology producers: using more technology to increase their profitability, but, this would presuppose that the cost of this increased use of technology would continue to rise faster than their incomes. Marx[50] already stated that the main cause of the decreasing tendency of the rate of profit is technology, since this presupposes that the organic composition of capital increases, thus decreasing the average rate of profit by proportionally decreasing living labour in relation to preterite labour, embodied by the increased use of machinery.

6.2 Recommendations

Water is an essential resource in the arid and semi-arid zones of Mexico, as its contribution limits the country's agricultural production. However, in this study it was observed that the real price of water represented a very low percentage of the cost of production, which leads to an irrational use of water. Therefore, water productivity studies should evaluate not only physical productivity but also economic and social productivity, since water use should be equitable, sustainable and socially responsible. It is therefore recommended that this type of study be carried out not only on a single crop but on the general cropping patterns that make up an agricultural region in order to find suitable patterns per agricultural year that can reduce the pressure exerted on water resources.

[50] Marx, C. 1985. Capital, volume I, Fondo de Cultura Economica, Mexico, D. F. see the chapter on the formation of the rate of profit and its decreasing tendency.

LITERATURE CITED

Aldaya M. M., Niemeyer I, and Zarate E. (2011). Water and Globalisation: Challenges and opportunities for a better management of water resources. Revista Espanola de Estudios Agrosociales y Pesqueros, 2011; 230: 61-83.

Aldaya, M. M., Novo, F., & Llamas, M. (2010). Incorporating the water footprint and environmental water requirements into policy: reflections from the Donana Region (Spain). *Re-thinking Water and Food Security, CRC Press/Balkema, London,* 193-218.

Barrientos, D. L. Isabel, (2015). Productivity of water, soil, capital and labour in apple (*Malus domestica*) cultivation in Canatlan, Durango, Mexico. Undergraduate thesis. Universidad Autonoma Chapingo, Unidad Regional Universitaria de Zonas Aridas, Bermejillo, Durango, Mexico. See Table 5, page 30.

Boutraa, T. (2010). Improvement of water use efficiency in irrigated agriculture: a review. *Journal of Agronomy*, *9*(1), 1-8.

Caravantes, R. E. D., Pena, L. C. B., Cejudo, L. C. A., & Flores, E. S. (2013). Anthropogenic pressure on groundwater in Mexico: a geographic approach. *Investigaciones Geograficas, Boletm del Instituto de Geografia*, *2013*(82), 93-103.

ECLAC (1982). Econom^a Campesina y Agricultura Empresarial (Tipolog^a de Productores del Agro Mexicano). Mexico, Siglo XXI Editores

Chapagain, A. K., & Hoekstra, A. Y. (2004). Water footprints of nations, value of water research Report Series No. 16.

CONAGUA (2010). Registro Publico de Derechos de Agua, Comision Nacional del Agua. Retrieved October 15, 2010. Available at www.conagua.gob.mx

CONAGUA (2011a), Atlas del Agua en Mexico, SEMARNAT-Gobierno Federal, Mexico.

CONAGUA (2011b), Estad^sticas del Agua en Mexico, National Water Commission. Available at: www.conagua.gob.mx

CONAGUA and COLPOS. (2007). Plan director: union de asociaciones de usuarios de aguas subterránraneas del acuifero de Cuauhtemoc, Chihuahua, S de RL de IP de CV, Comision Nacional del Agua.

CONAGUA (2009). Actualizacion de la Disponibilidad Media Anual de Agua Subterranea Acuifero (0805) Cuauhtemoc, Estado De Chihuahua, Comision Nacional del Agua. Available at: http://www.conagua.gob.mx

Cook, R. L. (1998, June). International trends in the fresh fruit and vegetable sector. In *WCHR-World Conference on Horticultural Research 495* (pp. 143-156).

Cooper, P. J. M., Dimes, J., Rao, K. P. C., Shapiro, B., Shiferaw, B., & Twomlow, S. (2008). Coping better with current climatic variability in the rain-fed farming systems of sub-Saharan Africa: An essential first step in adapting to future climate change? *Agriculture, Ecosystems & Environment*, *126*(1), 24-35.

De Vries, F. P. (2001). Food Security? We Are Losing 1 Ground Fast! *Crop science: Progress and prospects*, 1.

Falkenmark, M. & Rockstrom, J. (2006). The New Blue and Green Water Paradigm: Breaking New Ground for Water Resources Planning and Management. *J. Water Resour. Plann. Manage*. 132(3), 129-132.

Falkenmark, M., & Rockstrom, J. (2004). *Balancing water for humans and nature: the new approach in ecohydrology*. Earthscan.

Fereres, E., Orgaz, F., & Gonzalez-Dugo, V. (2011). Reflections on food security under water scarcity. *Journal of Experimental Botany*, *62*(12), 4079-4086.

Fischer, R. A., Byerlee, D., & Edmeades, G. O. Can technology deliver on the yield challenge to 2050? 46p.

Gamez, A.B. (2014). The production of strawberry *(Fragaria vesca)* from DR-14 Rio Colorado, BC and DR-085 Celaya, Guanajuato and its Myrica footprint. Bachelor's thesis. Universidad Autonoma Chapingo. Unidad Regional Universitaria de Zonas Aridas, Bermejillo, Durango.

Guerrero-Prieto, V., Trevizo, G., Figueroa, C., Romo, A., Gardea, A., & Blanco, C. (2004). Identification of epiphytic yeasts obtained from apple [Malus sylvestris (L.) Mill. var. domestica (Borkh.) Mansf.] for postharvest biological control. *Revista Mexicana de Fitopatologia*, *22*(2), 223-230.

Harrison, P., Bruinsma, J., de Haen, H., Alexandratos, N.,

Schmidhuber, J., Bodeker, G., & Ottaviani, M. G. (2002). World agriculture: towards 2015/2030. *Online, http://www. fao. org/documents*.

Hoekstra Arjen Y., Ashok K. Chapagain, Maite M. Aldaya and Mesfin M. Mekonnen. Chapagain, Maite M. Aldaya and Mesfin M. Mekonnen (2011). The Water Footprint Assessment Manual Setting the Global Standard. Earthscan Ltd, Dunstan House, 14a St Cross Street, London. ISBN: 978-1-84971-279-8.

INEGI. 2010. Municipal Geostatistical Framework 2010, version 5.0.

Jury, W. A., & Vaux, H. (2005). The role of science in solving the world's emerging water problems. *Proceedings of the National Academy of Sciences of the United States of America*, *102*(44), 15715-15720.

Marx, C. (1985). El Capital, vol. I, Fondo de Cultura Economica, Mexico, D. F. see the chapter on the formation of the rate of profit and its decreasing tendency.

Mata, E., O. A. (2004). Population Dynamics and Biological Management of Nematodes Associated with Apple (Pyrus malus L.) Crops, in two localities in Cuauhtemoc, Chihuahua. Bachelor Thesis. Universidad Autonoma Agraria "Antonio Narro" Division De Agronom^a. Buenavista, Saltillo, Coahuila. Mexico. March 2004

Molden D. (2007). *Water for food, water for life: a comprehensive assessment of water management in agriculture. London: Earthscan; Colombo: International Water Management Institute; 2007.*

Nature 2010. Can science feed the world? 2010. http://www.nature.com/news/specials/food/index.html. Accessed July 30, 2015.

UN, (2014). United Nations. Department of Economic and Social Affairs, http://www.un.org/es/html

Passioura, J. (2006). Increasing crop productivity when water is scarce-from breeding to field management. *Agricultural water management*, *80*(1), 176-196.

Pena, A. G., & Alatorre, L. C. Evaluation of groundwater abstractions by indirect methods in the Cuauhtemoc region, Chihuahua, Mexico: applying remote sensing and GIS. *Latin*

American Journal of Natural Resources, 9 (1): 141149.
State Development Plan 2004-2010. Chihuahua State Government. pp. 99-101.
Plan estatal de desarrollo 2010 -2016 (2010). Government of the State of Chihuahua, pp. 160-167.
Pontificia Universidad Catolica del Ecuador. Econom^a y Finanzas. http://www.puce.edu.ee/economia/efi
Postel, S. L. (2000). Entering an era of water scarcity: the challenges ahead. *Ecological Applications*, *10*(4), 941-948.
Quezada, R. A. P., Franco, P. O., Alvarez, J. P. A., & Sanchez, N. C. (2009). Apple tree productivity and growth under controlled deficit irrigation. *Terra Latinoamericana*, *27*(4), 337-343.
Ramkez-Legarreta, M. R., Jacobo-Cuellar, J. L., Avila-Marioni, M. R., & Parra-Quezada, R. A. (2006). Crop losses, production efficiency and profitability of apple orchards with varying degrees of technification in Chihuahua, Mexico. *Revista Fitotecnia Mexicana*, *29*(3), 215-222.
Salmoral, G., Dumont, A., Aldaya, M. M., Rodriguez-Casado, R., Garrido, A., & Llamas, M. R. (2012). *Analysis of the extended water footprint of the Guadalquivir river basin*. Marcelino Botin Foundation.
UACH, (2007). Evaluacion de Alianza para el Campo de los Sistemas Producto Fruttcolas en el Estado de Chihuahua. Universidad Autonoma de Chihuahua, 26-37pp.
UN-Water, (2012). The United Nations World Water Development: Managing Water under uncertainty and risk. World Water Assessment Programmed (WWAP). Report 4. Unesco, Paris, France.861p.

Printed by Books on Demand GmbH, Norderstedt / Germany